Landscaping with
Native Plants of Minnesota

Text and photography by
Lynn M. Steiner

Voyageur Press

Edited by Michael Dregni
Designed by Maria Friedrich
Printed in China

05   06   07   08   09   5   4   3   2   1

Library of Congress Cataloging-in-Publication Data

Steiner, Lynn M., 1958-
  Landscaping with native plants of Minnesota / by Lynn M. Steiner.
     p. cm.
  Includes bibliographical references and index.
  ISBN 0-89658-650-2 (pbk. : alk. paper)
  1. Native plant gardening—Minnesota.
  2. Native plants for cultivation—Minnesota.
  3. Landscape gardening—Minnesota. I. Title.
  SB439.24.M65S74 2005
  635.9'51776—dc22
                          2004023540

Published by Voyageur Press, Inc.
123 North Second Street, P.O. Box 338
Stillwater, MN 55082 U.S.A.
651-430-2210, fax 651-430-2211
books@voyageurpress.com
www.voyageurpress.com

On the cover: It may have the look of a native prairie planting, but this grouping of *Liatris pycnostachya* (great blazing star), *Ratibida pinnata* (gray-headed coneflower), *Rudbeckia hirta* (black-eyed Susan), *Andropogon gerardii* (big bluestem), and *Schizachyrium scoparium* (little bluestem) is part of a small backyard in the heart of Saint Paul. The garden was designed and installed by landscape designer and native-plant enthusiast Erik Olsen for his parents, Robert and Marlene Olsen.

On the title page: *Ratibida pinnata* (gray-headed coneflower) provides a nectar source for butterflies as well as seeds for birds.

On the contents page: *Aguilegia canadensis* (Canada columbine) and *Antennaria plantaginifolia* (pussytoes)

Dedication

This book is dedicated to the entire native-plant community of Minnesota, including the advocates and organizations; designers, nurseries, and garden centers; and home gardeners, who gave me their full support in writing this book. Thank you for your help and for all you do to preserve and promote the state's native treasures.

Acknowledgments

There are many people who have generously offered their help and shared their landscapes, without whom this book would not have been possible.

I'd especially like to thank the gardeners who allowed me to profile their landscapes in the Gallery of Gardens section: Barbara and Don Pederson, Pat and Bob Angleson, Peggy and Wayne Willenberg, Robert and Marlene Olsen, Mary and Dick Stanley, and Phil Friedlund and Lisa Isenberg. Their willingness to open their gardens to me and my camera is greatly appreciated.

Thanks also to the other people who allowed me to photograph their gardens, homes, and places of business, including Diane Hilscher, Fred and Sharon Remund, Barb Staub and Don Mitchell, Claire Olsen, Marge Hols, Sue Price, Deb Revier, Paul and Susan Damon, Doug and Sue Law, Carla Henry, Richard and Olive Zoller, Outback Nursery, and Landscape Alternatives.

For their help in locating the landscapes to use in the book, I'd like to thank the staff of Prairie Restorations, Tom Tenant and Erik Olsen of Outback Nursery, Diane Hilscher of Hilscher Design and Ecology, and Char Menzel.

For help with plant information and locating plants, I'd like to thank Nancy Rose, University of Minnesota Extension Service; Fred Rozumalski, Barr Engineering; Rick Sandager, Abrahamson's Nurseries; Roy Robinson, Landscape Alternatives; and David Stevenson, Minnesota Landscape Arboretum.

I'd like to recognize John R. Tester's *Minnesota's Natural Heritage: An Ecological Perspective* and the Minnesota Department of Natural Resource's website, www.dnr.state.mn, for the valuable information each provided on Minnesota's natural heritage and native plant communities. I'd also like to recognize the Minnesota Landscape Arboretum, where I was able to photograph and study many of the native plants found in this book.

I'd like to thank John Whitman, Cole Burrell, and Mike Heger for their help in determining how to proceed with the writing of this book.

Lastly, I'd like to thank my family and friends for their support of this project.

Contents

Landscaping with
Native Plants of Minnesota

Why a book on landscaping with native plants of Minnesota? Mainly because it hasn't been done and there is a definite need. There are many good field guides and reference books for identifying and learning about native plants. There are also several good books about gardening in Minnesota. This book combines these two approaches—identifying Minnesota's native plants and plant communities and demonstrating how to use them effectively in a typical home landscape.

I hope to dispel the many misconceptions people have about native plants, such as their being weedy, hard to grow, difficult to purchase, and generally inappropriate for landscape situations. I also hope to show you the many possibilities available in our native flora so you can choose plants that are both appealing and well adapted

to the climate and soils of your landscape and well suited to your lifestyle.

My first task was determining which plants are "native." I followed the distinction used by many experts and based my selection on what was growing here naturally before European settlement. For this information, I turned to *Vascular Plants of Minnesota* by Gerald B. Ownbey and Thomas Morley, and to the Minnesota Department of Natural Resources website on state vascular plants prepared by Welby R. Smith. I have not included plants that have become naturalized—often labeled "wildflowers"—and aren't indigenous to any part of the state.

My initial list of native plants was far too long, so I had to make difficult choices about which plants to leave out. I based my decision on each plant's ability to adapt to cultivation, suitability for various landscape situations, and availability at local nurseries or through mail-order sources. One of the true joys of working with native plants is observing how they match the rhythms of the seasons, so I looked for plants that provide a succession of interest year round. I also tried to include a variety of sun and shade plants and a cross section of plants native to the state's three major biomes. To the many wonderful plants that didn't make the cut, I apologize. I eliminated plants that are difficult to bring into the garden because of demanding soil or cultural requirements or are difficult to find.

Space limitations also prevented me from including detailed propagation information for each species. There are many good references out there on native-plant propagation—especially for wildflowers. If you are interested in this fascinating aspect of gardening with native plants, I encourage you to read *Restoring the Tallgrass Prairie: An Illustrated Manual for Iowa and the Upper Midwest* by Shirley Shirley; *The New England Wild Flower Society Guide to Growing and Propagating Wildflowers of the United States and Canada* by William Cullina; or *Native Trees, Shrubs, and Vines: A Guide to Using, Growing, and Propagating North American Woody Plants* by William Cullina.

What else is included in this book? You'll find an overview of Minnesota's natural heritage, an understanding of which is crucial before attempting any type of native landscaping. Still, nothing you read here will be more informative than what you can learn from nature itself: be sure to take lots of walks in Minnesota's many parks and nature areas for inspiration.

In this book you'll also find basic gardening information tailored to native plants. You'll learn what level of native-plant landscaping is right for you and get valuable information on the process of designing a natural garden that fits your lifestyle. You'll also find lots of plant lists for specific styles of gardens.

In the Gallery of Gardens section, you'll be inspired by what your fellow Minnesota gardeners have done with native plants in their own landscapes, including a prairie restoration, a suburban woodland garden, and a garden for wildlife.

The Native Plant Profiles section includes comprehensive descriptions of some 350 species of flowers and groundcovers, trees, shrubs, vines, evergreens, grasses, and ferns native to Minnesota, as well as information on planting, maintenance, and landscape uses for each plant.

Late July color comes from *Heliopsis helianthoides* (oxeye), *Monarda fistulosa* (wild bergamot), *Verbena stricta* (hoary vervain), and *Artemisia ludoviciana* (prairie sage).

Facing Page: Native plants replace customary nonnative shrubs in this foundation planting.

Chapter 1

Understanding
Native Plants

What They Are and How to Use Them

Just what defines a native plant has been debated for many years by many people. A widely accepted definition—and the one used in this book—classifies native plants as those species that grew in an area before European settlement—about the mid-1800s in the Midwest. By and large, Native Americans lived in harmony with the plants and animals of an area without endangering the natural ecosystems. European settlers, on the other hand, had a major impact on the landscape as they cut down large stands of trees, plowed up acres of prairies, suppressed natural fires, and introduced plants from their homelands and other parts of this "new" continent.

Unlike most introduced plants, a native plant fully integrates itself into a biotic community, establishing complex relationships with other local plants and animals. Not only does a native plant depend on the organisms with which it has evolved, but the other organisms also depend on it, creating a true web of life. This natural system of checks and balances ensures that native plants seldom grow out of control in their natural habitats.

"Wildflower" is a commonly used term, but it does not necessarily mean a native plant, since not all wildflowers are native to an area. Wildflowers include introduced plants that have escaped cultivation and grow wild in areas. Examples are Queen Anne's lace (*Daucus carota*) and chicory (*Chicorium intybus*), two common roadside plants, neither of which is native to any area of the United States.

Introducing new plants is not always a bad thing. Where would we be without tomatoes, potatoes, and wheat? And, it's hard to find fault with introduced plants as charming and well behaved as lilacs and hostas. However, experience has taught us that the introduction of nonnative plants into an ecosystem is a delicate operation that should be undertaken with care. More and more, we are finding that plants that evolved in other countries, or even other areas of this country, can become too comfortable in landscape situations and threaten native flora. A prime example is purple loosestrife, a European native propagated by nurseries and grown in gardens for years before it was realized that it aggressively invades natural wetlands, crowding out native plants. Buckthorn is another European native that has been widely used as a hedge. Today, it is the bane of any homeowner with a wooded plot and it is running rampant through native woodlands.

Classifying Native Plants

Before you can know and effectively use native plants, you must have a simple knowledge of plant taxonomy. The fundamental category in this book is the species, a group of genetically similar plants within a genus, a larger botanical division. Genus and species names are commonly Latin and italicized, with the genus name coming first in capital letters followed by the species in lower case. Learning Latin names can be frustrating, but it is important. Too many plants share the same or similar common names, and it's easy to end up with the wrong plant—one that may not even be native to your area. Latin names also offer clues on how to identify a plant. For example, knowing that *tomentosus* means "downy" and *laevis* means "smooth" will help you identify and remember what a plant looks like.

Within a species, there are also subspecies (abbreviated as "ssp.") and varieties ("var."). A subspecies has a characteristic that isn't quite different enough to make it a separate species. This characteristic may occur over a wide range or in a geographically isolated area.

Varieties have minor recognizable variations from the species, such as flower size or leaf color, but are not distinct enough to be labeled subspecies. An example is found in *Cypripedium calceolus* (yellow lady's slipper), which is further differentiated into var. *pubescens* (large yellow lady's slipper) and var. *parviflorum* (small yellow lady's slipper). The latter plant is shorter and has a slightly different flower shape and color, but without seeing the two side by side, it can be difficult to tell which one you are looking at.

As native plants become more popular, many horticulturally selected cultivated varieties are being introduced. These "cultivars" are usually chosen for certain characteristics such as larger or double flowers, leaf color, compact growth, or flower color, and are propagated by nurseries to maintain the trait. In most cases, these cultivars retain most of the characteristics of the native species and are fine choices for most landscape use. However, if you are doing restoration work, you will want to stick with the species or even the subspecies or variety native to your area to maintain the true genetic diversity you'll only get from the native species.

For most plants, species is the final classification. However, some plants are divided further into varieties. When you see them side by side, you can see that the *Cypripedium calceolus* var. *pubescens* (large yellow lady's slipper) at left and var. *parviflorum* (small yellow lady's-slipper) at right have minor size and color differences, thus the further differentiation into varieties. Some botanists feel that the differences are distinct enough to classify *C. pubescens* as a separate species.

While the native pink-flowered *Physostegia virginiana* (obedient plant) is a beautiful flower suitable for naturalizing and prairie plantings, it can be too aggressive for many landscapes. 'Miss Manners', a white cultivar, is less aggressive and better suited to garden use.

The Bad Guys: Invasive and Weedy Introduced Plants

While many introduced plants are well-behaved, beautiful additions to gardens, some become harmful invaders of local habitats. You may be surprised by how many common landscape plants have the potential to become invasive and weedy when grown in conditions that promote rampant growth, such as tended garden beds or abandoned and neglected sites where native plants are no longer prevalent. Here are some of the many plants that have been turning up on invasive-plant lists in recent years:

Acer ginnala (Amur maple)
Acer platanoides (Norway maple)
Aegopodium podagraria (goutweed)
Ailanthus altissima (tree of heaven)
Ampelopsis brevipedunculata (porcelain berry)
Berberis thunbergii (Japanese barberry)
Berberis vulgaris (common barberry)
Butomus umbellatus (flowering rush)
Campanula rapunculoides (creeping bellflower)
Caragana arborescens (Siberian peashrub)
Celastrus orbiculatus (Oriental bittersweet)
Coronilla varia (crown vetch)
Daucus carota (Queen Anne's lace)

Elaeagnus angustifolia (Russian olive)
Euonymus alatus (winged euonymus)
Euonymus europaeus (European euonymus)
Euonymus fortunei (winter creeper euonymus)
Euphorbia esula (leafy spurge)
Glechoma hederacea (creeping Charlie)
Hesperis matronalis (dame's rocket)
Hieracium aurantiacum [*Pilosella aurantiaca*] (orange hawkweed)
Iris pseudacorus (yellow flag)
Leucanthemum vulgare (oxeye daisy)
Ligustrum vulgare (common privet)
Lonicera japonica (Japanese honeysuckle)
Lonicera tatarica (Tatarian honeysuckle)

Lotus corniculatus (bird's foot trefoil)
Lythrum salicaria (purple loosestrife)
Miscanthus sinensis (maiden grass)
Phalaris arundinacea (reed canary grass)
Polygonum japonicum [*P. cuspidatum*] (Japanese knotweed)
Rhamnus cathartica (common buckthorn)
Robinia pseudoacacia (black locust)
Rosa multiflora (multiflora rose)
Saponaria officinalis (soapwort, bouncing bet)
Sorbus aucuparia (European mountain ash)
Tanacetum vulgare (common tansy)
Ulmus pumila (Siberian elm)
Vinca minor (common periwinkle)

Source: Plant Conservation Alliance's Alien Plant Working Group: http://www.nps.gov/plants/alien

The term "wildflower" can be a misnomer when referring to native plants as not all plants that grow well in a region are native there. If you see a proliferation of one species—such as the oxeye daisy (*Chrysanthemum vulgare*), a European native—taking over lawns and roadsides, chances are it's an introduced plant displacing native species.

Many aggressive nonnative plants are introduced as garden plants because they are showy and easy to grow. *Campanula rapunculoides* (creeping bellflower) is an example of a European native that has become a prolific garden weed. Its thick, tuberlike rhizome makes it difficult to eradicate once it becomes established.

Benefits of Native Plants

There are many reasons to use native plants, some more tangible than others. For many gardeners, the initial attraction comes from native plants' reputation of being lower maintenance than a manicured lawn and exotic shrubs. For the most part this is true—provided native plants are given landscape situations that match their cultural requirements. Because they have evolved and adapted to their surroundings, native plants tend to be tolerant of tough conditions such as drought and poor soil. Native plants are better adapted to local climatic conditions and better able to resist the effects of native insects and diseases. Their reduced maintenance results in less dependence on fossil fuels and reduced noise pollution from lawn mowers and other types of equipment.

The less tangible—but possibly more important—side of using native plants is the connection you make with nature. Gardening with natives instills an understanding of our natural world—its cycles, changes, and history. Communing with nature has a positive, healing effect on human beings. Learning how to work with instead of against nature will do wonders for your spiritual health.

By observing native plants throughout the year, a gardener gains insight into seasonal rhythms and life cycles. You will experience intellectual rewards that are somehow missing if you only grow petunias or marigolds.

Gardening with native plants will help you create a sense of place rather than just a cookie-cutter landscape. Your yard will be unique among the long line of mown grass and clipped shrubs in your neighborhood. A native-plant landscape will blend into the natural surroundings better that those planted with introduced species, and you will get an enormous sense of satisfaction from helping reestablish what once grew naturally in your area. You will see an increase in wildlife, including birds, butterflies, and pollinating insects, making your garden a livelier place.

On a broader scale, using native plants helps preserve the natural heritage of an area. Genetic diversity promotes the mixing of genes to form new combinations, the key to adaptability and survival of all life. Once a species becomes extinct, it is gone forever, as are its genes and any future contribution that it might have made.

Misconceptions about Native Plants

Despite the increased interest and promotion of native plants, many people still hesitate to use them for one reason or another. Here are some of the common misconceptions about using native plants.

Native plants are colorless and dull
The belief that native plants are drab or uninteresting is based in ignorance. Once you learn about the wide variety of natives and how to use them properly, you will discover that they have much to offer, not only colorful flowers but also interesting textures, colorful fruits, and year-round interest. They may not all be as bright and showy as a lot of introduced plants, but their subtle beauty can be just as effective in landscaping.

Native plants cause allergies
The truth is, most native plants are insect pollinated rather than wind pollinated. Kentucky blue grass has the potential to produce more allergens than any native plant.

Native plants are invasive
Most aggressively invasive plants are imported from other countries or another part of the United States. Keep in mind that any plant can become invasive if it is given the right conditions—a site more conducive to rampant growth than its preferred habitat.

Native plants are hard to grow
The misconception that native plants are hard to grow comes from the fact that many of them have evolved in a rather specific habitat. Once you learn about the different plant communities and their soil and sunlight requirements and determine which plants are best for your conditions, you will find that most native plants are easier to grow than their cultivated counterparts.

Native plants are messy
Nature is "messy." It's full of fallen logs, recycling plant parts, and plants that weave together rather than lay out in straight lines. Once you understand and appreciate this, native plants will no longer appear unattractive. When given proper conditions and room to grow, most native plants produce larger and better flowers than their wild counterparts. There are many things you can do to make a native landscape look neater, such as incorporating small patches of lawn grasses, creating paths and neat edges, and cutting back certain plants when they are done blooming.

Native plants are hard to find
Once you learn which plants are native, you will be surprised how many are available at local nurseries. In every part of the country, you will find nurseries that specialize in native plants, and many of them offer mail order.

Native Plants in the Landscape

Basically there are two ways to use native plants in a landscape. On the extreme side, you can grow only plants that were found in your area before European settlement. This is a wonderful way to preserve and enjoy the beauty of individual plants, and also to preserve entire ecosystems or plant communities. Restoring a tall-grass prairie or creating an authentic deciduous woodland habitat are wonderful ways to create pure stands of native plants.

For most people, however, using only native plants is not practical. Because our landscapes have been altered so much by human activity, it is difficult to go back to the point where you can successfully grow only native plants without investing quite a bit of time and effort in plant eradication and site preparation. If you truly want to establish a pure stand of plants that once grew naturally in your area, you should get help from a professional specializing in native-plant restorations.

For most gardeners, a more practical way to use native plants is to integrate natives with nonnative, more traditional landscape plants that have proven to be nonaggressive and adaptable to your area. You may already be doing this without realizing it. If your mixed border includes liatris, butterfly weed, and black-eyed Susans, or if your shade garden is home to wild ginger, pagoda dogwood, and maidenhair ferns, you are already well on your way to using native plants. You don't have to give up some of the well-behaved nonnative plants you love, such as spring bulbs, hostas, and rhododendrons. Most natives are adaptable and willing to coexist with nonnatives. Be warned, however: once you have discovered the subtle beauty that natives bring to your landscape, you may well become one of the many gardeners choosing to grow more and more of these fascinating plants.

The bottom line is, there is really no right or wrong way to use native plants, as long as it brings you pleasure. As with any type of gardening, your landscape should reflect your own preferences for color, style, and plants. If you have a colonial-style house in the suburbs and prefer a formal entryway with clipped hedges, there are native plants that will fit the bill. If you're a plant lover who can't resist the hodgepodge of a collector's garden, that's fine too. No matter how many natives you use or in which way you use them, you will be helping to counteract the tragedy of habitat destruction and reduction in native-plant populations occurring around the world. And, on a more personal note, you won't have to leave home to enjoy nature. It will be right at your doorstep.

No single gardener can solve all the environmental problems of the world, but by growing native plants, you can help preserve and promote the natural ecosystem of your small part of the globe. Like many native plants, *Geranium maculatum* (wild geranium) may not be as ostentatious as its cultivated cousins, but its subtle beauty, adaptability, and low maintenance make it easy to use in many landscape situations.

Although it takes several years to establish, if you have the space and resources, a prairie restoration is a wonderful way to experience the wonders of a natural plant community.

Native Plant Conservation

Once you've been convinced of the benefits of growing native plants, it's time to temper that recommendation by saying it's important to use them properly and responsibly.

Native plant gardeners face moral and ethical considerations that most gardeners do not. You must be sure that the native plants you buy are propagated by a nursery and not collected in the wild. You want to purchase plants that were "nursery propagated," not just "nursery grown." Reputable nurseries will readily volunteer information on the origin of their plants, so evasiveness or ambiguous answers from nursery owners should trigger caution.

Plants growing in their native habitats should never be dug up for garden use unless the plants are facing imminent destruction from development. It is always preferable to try to preserve or restore a natural habitat rather than destroy it, but sometimes this just isn't possible. If you have permission to collect seeds from a stand of native plants, take only what you need. Collect only a few seeds from several plants in the stand; never take all of the seeds from one plant. Do not collect underground plant parts. Collecting must never endanger a plant population.

The rapid destruction of native habitats in the last century means many native plants and animals are threatened with extinction. You should become aware of these plants, since in many states it is illegal to gather, take, buy, or sell plants listed as endangered or threatened. A safeguard for endangered native species is the Federal Endangered Species Act of 1973. This law applies only to federal lands, however. Protection of endangered plants on other public and private lands is left up to individual states, and each state has its own list of endangered, threatened, and special-concern plants.

A complete habitat restoration is not practical for most homeowners. Luckily, many native plants adapt readily to traditional landscape use. Here *Echinacea purpurea* (purple coneflower), *Ratibida pinnata* (gray-headed coneflower), *Eupatorium purpureum* (Joe-pye weed), *Liatris pycnostachya* (great blazing star), and native grasses help make the transition from the water's edge to a more formal part of this landscape.

How to Select Propagated Over Collected Native Plants

For the conservation of native plant species, it's important the native plants you purchase at a nursery are propagated rather than collected from the wild. Despite assurances from nursery owners, you should trust your own eyes as well. Here are some signs that plants may be collected instead of propagated legitimately by a nursery:

Poor or abnormal color
Sparse foliage
Weak stems or wilted leaves
Legginess
Large size or obvious maturity, especially in
    slow-growing plants

Off-centered potting
More than one species in the same pot
Large stones in the soil
Different soil types in the same pot
Compacted clay rather than uniformly
    textured potting soil

Hydrastis canadensis (golden seal) is a fascinating native plant with interesting leaves and showy flowers and fruits. Unfortunately, its reputed medicinal qualities have lead to over-collecting in the wild, and it is now classified as a threatened or endangered species in several states. Be sure to only purchase nursery-propagated plants if you want to include this gem in your landscape.

Plant collecting can be devastating to native plant communities. Always ask the nursery or supplier where they got their plants. If they are hesitant or evasive about their sources, take your business elsewhere.

Chapter 2

Minnesota's Natural Plant Life

Learning from the Natural World

The key to creating a successful home landscape using native plants is to understand the natural plant communities in your area. Nature is truly the best garden designer, and you will never go wrong if you attempt to imitate it.

For gardeners in Minnesota, the natural world has provided many options. Minnesota is the meeting point for three North American biomes, a rare occurrence in most states. Minnesota flora represents the western limits of the vast eastern forest flora and the northern and eastern limits of the flora of the prairies and plains in the continental United States. As a result, the state's natural vegetation shows up on the map in roughly three parallel bands that cross the state diagonally.

The first band, covering the southern and western parts of the state, is the tall-grass prairie, with its rolling waves of grasses and myriad flowers. In the southeast and extending in a narrow corridor northwestward nearly to the Canadian border are deciduous forests composed mainly of sugar maples, basswoods, and elms. This ecosystem includes the area known to early settlers as the Big Woods. (Between the prairie and the Big Woods lies a band of oak savanna, an ecosystem often set apart as a separate biome, but here it is included as a subhabitat within the prairie.) The northeastern third of the state is a mosaic of stands of white and red pines interspersed with spruces, firs, aspens, and birches making up the northern coniferous forest. Minnesota is also home to many different water features, ranging from large lakes to bogs, each of which takes on different characteristics depending on which plant zone it is located in.

Some of Minnesota's native plants evolved in only one or two of the three general plant communities described above, and some are native throughout the state. Obviously, plants do not recognize political boundaries such as state lines, and these plant communities spill over into neighboring states and provinces. As a result, the plant communities of southeastern Minnesota are more like the plant communities of southwestern Wisconsin and northeastern Iowa than the plant communities of northwestern Minnesota.

There is one plant, however, endemic to Minnesota. This is the dwarf, or Minnesota, trout lily (*Erythronium propullans*), which grows in mature maple-basswood forests and adjoining floodplains in Rice and Goodhue Counties and nowhere else in the world. Because it is so rare, the federal government classified it as an endangered species in 1986.

Major Vegetative Regions of Minnesota

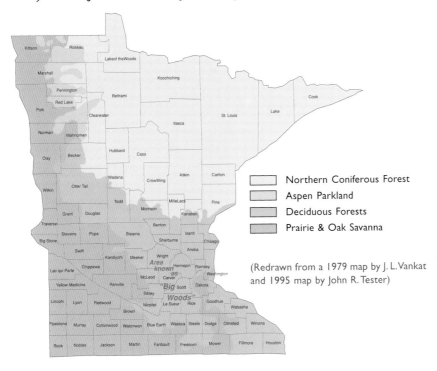

Northern Coniferous Forest
Aspen Parkland
Deciduous Forests
Prairie & Oak Savanna

(Redrawn from a 1979 map by J. L. Vankat and 1995 map by John R. Tester)

Major Ecological Regions of the United States

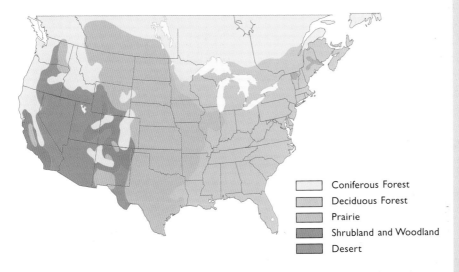

Coniferous Forest
Deciduous Forest
Prairie
Shrubland and Woodland
Desert

Facing page: A successful native landscape starts with a good understanding of your area's ecosystems and the plants that grow in them. Sharon and Fred Remund of Marine, Minnesota, have successfully interpreted the natural plant community found several miles north of them in William O'Brien State Park. They've allowed *Caltha palustris* (marsh marigold) and *Symplocarpus foetidus* (skunk cabbage) to grow along their stream, and the rustic wooden bridge mimics the natural fallen log.

Minnesota's Geography and Climate

Minnesota's area is 86,943 square miles; 7,326 square miles are covered by water if you include the state's portion of Lake Superior. Most of the state is characterized by gently rolling hills. The northwestern corner contains broad expanses of flat land, and the northeast contains rugged hills that drop to the shoreline of Lake Superior, Minnesota's coastline.

The mean elevation of Minnesota is 1,200 feet above sea level. The state's highest and lowest elevations are found in the Arrowhead Region. Lake Superior's shore is the lowest elevation at 602 feet above sea level; Eagle Mountain is the highest point at 2,301 feet.

Average temperatures are about 2 to 3 degrees Fahrenheit cooler for every 100 miles of northward travel. The average January temperature is about 0 degrees in the northwest and about 14 degrees in the south. The average temperature in July is about 74 degrees in the south and as low as 60 degrees in the far northeast. There are bog areas in northern Minnesota that have reported freezing temperatures every month of the year.

The frost-free growing season generally ranges from about the second week of May in the south and ends during the first week of October. The growing season starts about June 1 in the north and ends about mid-September.

Annual precipitation varies from 18 to 20 inches in the northwest to 32 inches or more in the southeast. A little more than two-thirds of the annual precipitation falls from May through September (the growing season). The average annual snowfall in Minnesota varies from 36 inches in the southwest to more than 70 inches along Lake Superior.

Source: State Climatology Working Group: http://www.climate.umn.edu

Tall-Grass Prairie

Of the three major biomes found in Minnesota, the tall-grass prairie has been most devastated by humankind. Prior to European settlement, more than 18 million acres of tall-grass prairie stretched across Minnesota. Today less than 1 percent of Minnesota native prairie remains, and most of the prairie wetlands have been drained. About 150,000 acres of native prairie remain, with about 48,000 acres protected by the DNR, the Nature Conservancy, and other agencies.

The main reason for this mass destruction of habitat is that Minnesota's native prairies contain some of the most fertile soils on Earth. Once early settlers discovered this soil, they set about plowing up acres and acres without a thought to the native flora and fauna. It's not surprising that the prairie biome is home to a greater number of endangered, threatened, or special-concern animal and plant species than either of the forest biomes. Unfortunately, once the prairie sod has been broken and the soil farmed for several years, it can take a century or more for a tall-grass prairie to reestablish itself.

The boundary between the tall-grass prairie and its neighboring ecosystem, the deciduous forest, was constantly in flux, creating the savanna, an area with scattered trees and large open areas of prairie vegetation. Rainfall amounts and the occurrence of fires prevented woody species from encroaching onto the prairie. The climate of the northeastern parts of the tall-grass prairie was moist enough that trees and shrubs could encroach into grasslands were it not for the periodic fires that destroyed most woody plants.

Rainfall amounts not only kept woody species out, but also determined what prairie species grew in an area. Root systems of prairie plants are variable, ranging from deep tap roots with fine rootlets that can tap deep subsoil reserves in times of drought to shallow, dense root systems that rely on light rains that only penetrate a few inches into the soil. These variations in soil moisture and the plant species adapted to them have led to the classification of prairies into three main types: mesic, dry, and wet. Mesic prairies, having moderate moisture, are the most common native grassland ecosystem in Minnesota.

Most prairie plants evolved to grow from their base or just below the soil surface, a trait that allowed them to not only survive repeated fires but to actually benefit from them. Aboveground plant parts were killed by fire, recycling nutrients back into the soil and allowing new shoots to grow readily from the plant's crown. With the litter layer burned off by fire, plants could grow quickly in the ample sunlight and fertile soil. This method of growing also allowed prairie plants to withstand repeated trampling and grazing by prairie animals, along with the high winds common on the prairie.

Minnesota's tall-grass prairies were home to a mix of forbs and grasses, with grasses and sedges comprising 75 to 80 percent of the biomass.

Prairie in the Landscape

Of Minnesota's three main plant biomes, the plants of the tall-grass prairie are probably the most adaptable to general landscape use. Many have showy flowers and have evolved to withstand the hot, sunny conditions typical of many perennial gardens. The earliest species generally start blooming in late spring, with later species continuing their show right up to frost. Many flowers have persistent seed heads offering winter interest, and prairie grasses remain showy until late winter.

It is possible to create the look of a prairie without doing a full-fledged restoration. Select a sunny part of your landscape with few or no large trees. An occasional oak or maple will a give savanna look. Select a mixture of prairie forbs and grasses based on your soil type: mesic plants for average soil; dry-prairie plants for well-drained, sandy soil (page 20); and wet-prairie plants for heavier soil (page 53).

For a natural look, be sure to use plenty of native grasses. Their deep roots aerate soil, improve drainage, and contribute substantial amounts of humus when they die back. Grasses also support stems of wildflowers and make your prairie or savanna look more natural.

Embellish your prairie landscape with paths, ornaments, and furniture that match its informal style. Incorporate broad and winding gravel, wood-chip, or mown paths, and place casual benches made of wood, rusted metal, or sawn logs along the paths. For ornament in prairie landscapes, have fun using "junk" salvaged from farmyards, which is so popular at garden centers and gift shops. Bird and butterfly houses will provide shelter for all the fluttering friends who'll be visiting, and a birdbath or other source of water will also be appreciated.

Savannas were found along the border of the tall-grass prairie and the deciduous forests. In contrast to prairies, savannas were home to widely spaced, fire-resistant trees and occasional shrubs such as *Rhus glabra* (smooth sumac) and *Amorpha canescens* (leadplant).

Include several benches in your landscape where you can stop and enjoy what you have created. This rustic wooden bench is well suited to the informality of this large prairie garden.

Native Plants of the Tall-Grass Prairie

The tall-grass prairie was once home to more than 900 native species. Here are some common plants of Minnesota's tall-grass prairies and savannas that adapt well to cultivation.

Grasses

	Mesic Prairie	Dry Prairie	Savanna
Andropogon gerardii (big bluestem)	x	x	x
Bouteloua curtipendula (side-oats grama)		x	x
Elymus canadensis (nodding wild rye)	x		
Panicum virgatum (switch grass)		x	
Schizachyrium scoparium (little bluestem)			x
Sorghastrum nutans (Indian grass)	x	x	x
Sporobolus heterolepis (prairie dropseed)	x	x	x

Forbs

	Mesic Prairie	Dry Prairie	Savanna
Allium cernuum (nodding wild onion)		x	x
Antennaria species (pussytoes)		x	
Asclepias tuberosa (butterfly weed)		x	
Asclepias verticillata (whorled milkweed)		x	
Asclepias viridiflora (green milkweed)		x	
Aster cordifolius (heart-leaved aster)			x
Aster ericoides (heath aster)	x	x	
Aster laevis (smooth aster)	x	x	x
Aster novae-angliae (New England aster)		x	
Aster oolentangiensis (sky-blue aster)	x	x	
Baptisia lactea (white wild indigo)		x	x
Campanula rotundifolia (harebell)		x	
Coreopsis palmata (stiff tickseed)		x	x
Dalea candida (white prairie clover)	x	x	
Dalea purpurea (purple prairie clover)		x	
Desmodium canadense (showy tick trefoil)	x		x
Dodecatheon media (prairie shooting star)	x	x	x
Echinacea angustifolia (narrow-leaved purple coneflower)			x
Eryngium yuccifolium (rattlesnake master)		x	x
Eupatorium purpureum (sweet Joe-pye weed)			x
Euphorbia corollata (flowering spurge)	x	x	
Gentiana andrewsii (bottle gentian)	x		
Geranium maculatum (wild geranium)			x
Geum triflorum (prairie smoke)			x
Heliopsis helianthoides (oxeye)	x	x	x
Heuchera richardsonii (alumroot)	x	x	x
Liatris aspera (rough blazing star)			x
Liatris pycnostachya (great blazing star)	x		
Lilium michiganense (Michigan lily)	x		x
Lobelia siphilitica (great lobelia)	x		

Forbs (continued)

	Mesic Prairie	Dry Prairie	Savanna
Lupinus perennis (wild lupine)		x	
Monarda fistulosa (wild bergamot)	x		
Phlox pilosa (prairie phlox)	x	x	x
Physostegia virginiana (obedient plant)			x
Pulsatilla patens (pasque flower)			x
Pycnanthemum virginianum (Virginia mountain mint)		x	
Rudbeckia hirta (black-eyed Susan)	x	x	
Silphium laciniatum (compass plant)	x	x	
Sisyrinchium angustifolium (blue-eyed grass)	x		
Solidago flexicaulis (zigzag goldenrod)			x
Solidago nemoralis (gray goldenrod)		x	
Solidago rigida (stiff goldenrod)		x	x
Solidago speciosa (showy goldenrod)	x	x	
Thalictrum dasycarpum (tall meadow rue)	x		
Tradescantia ohiensis (Ohio spiderwort)	x	x	x
Verbena stricta (hoary vervain)			x
Vernonia fasciculata (bunched ironweed)		x	
Veronicastrum virginicum (Culver's root)	x		x
Zizia aurea (golden alexanders)	x		x

Woody Plants

	Mesic Prairie	Dry Prairie	Savanna
Amelanchier arborea (downy serviceberry)			x
Amorpha canescens (leadplant)		x	x
Betula papyrifera (paper birch)			x
Ceanothus americanus (New Jersey tea)			x
Cornus racemosa (gray dogwood)			x
Cornus stolonifera (red-osier dogwood)			x
Corylus americana (American hazelnut)			x
Juniperus virginiana (eastern red cedar)			x
Prunus americana (wild plum)			x
Prunus pumila (sand cherry)			x
Prunus virginiana (chokecherry)			x
Quercus alba (white oak)			x
Quercus ellipsoidalis (northern pin oak)			x
Quercus macrocarpa (bur oak)			x
Rhus glabra (smooth sumac)			x
Rosa arkansana (prairie rose)			x
Symphoricarpos albus (snowberry)			x
Viburnum lentago (nannyberry)			x
Viburnum rafinesquianum (downy arrow wood)			x

Deciduous Forest

Deciduous forests existed mainly in a band stretching diagonally across Minnesota from northwest to southeast, with smaller patches in other areas of the state. The main band was not continuous, but rather a mosaic of forests, brushlands, oak woodlands, prairies, and savannas. This band developed because temperatures in the north restricted growth of many of the plants of this region and prairie fires and lack of rainfall kept them from encroaching into the prairie biome. Although fires occurred within this region, they were less common than on the western prairies, primarily due to irregular topography and the presence of lakes.

The most famous and possibly the most interesting of Minnesota's deciduous forests is the climax maple-basswood forest, also known fondly as the Big Woods. Once covering more than 2 million acres west and south of the Twin Cities, nearly all of the 3,000-square-mile Big Woods has been converted to farmland or developed. Isolated stands of deciduous forest still exist in parks, scientific and natural areas, and other lands owned by the DNR, the Nature Conservancy, or privately.

The maple-basswood forest is found on mesic sites with moderate soil moisture. Because the dense tree canopy of the deciduous forest permits so little light to reach the forest floor during summer, it is home to many spring ephemerals—herbaceous plants that bloom, produce seeds, and die back in May and early June before tree leaves are fully developed. Life is a race for most of these diminutive forest denizens, a case of the earliest bloomers getting the sunlight. Fully leafed-out deciduous trees plunge the forest floor into shade by early summer, blocking out as much as 95 percent of the available sunlight. Trout lilies, Dutchman's breeches, spring beauties, cut-leaved toothwort, and false rue anemone have evolved to emerge before trees leaf out and quickly convert early spring sunshine into food to be stored or used for growth and development of flowers and fruits. Other shade-tolerant wildflowers such as May apples, bloodroot, Jack-in-the-pulpit, wild ginger, hepaticas, and trilliums retain their leaves after the canopy emergence and ripen their fruits in midsummer. Maple and basswood saplings and shade-tolerant shrubs such as leatherwood, American hornbeam, ironwood, bitternut hickory, and pagoda dogwood dominate the forest's understory, or shrub layer.

There are other deciduous forest groupings within this biome distinguished by their soil moisture. Xeric forests, found on dry sites and featuring drought-tolerant species, are typically dominated by white, red, and black oaks. Lowland forests, found along floodplains and swamps, are adapted to the greatest extremes in moisture, ranging from spring flooding to summer drought. Canopy species vary widely. Floodplain forests include silver maple, cottonwood, black willow, American elm, green ash, and bur oak. Hardwood-swamp forests include black ash, paper birch, yellow birch, red maple, American elm, slippery elm, and green ash.

Big Woods State Park near Nerstrand, Minnesota, is a good example of the original "big woods" that once covered parts of south-central Minnesota.

St. Paul garden designer and writer Marge Hols turned a shady weed patch along the east side of her house into a woodland retreat reminiscent of a native deciduous woodland habitat.

Deciduous Forest in the Landscape

The key to successfully interpreting Minnesota's deciduous forest is to think in terms of layers. The top layer is the tree canopy, creating dense shade by early summer. The middle layer is made up of shade-tolerant small trees and large shrubs. The last layer, formed of spring ephemerals, ferns, and shade-tolerant groundcovers and flowers, creates the carpet of green that covers the forest floor.

If you are lucky enough to have a woodlot on your property, you are well on your way to having a woodland garden. Remove debris and most of the dead trees, both standing and fallen. Weed out undesirable trees and shrubs, including most hardwoods with trunks less than 6 inches or so in diameter, and buckthorn, poison ivy, and other exotics. After laying out winding paths, you can start adding shade-tolerant shrubs, groundcovers, flowers, and ferns.

If your property is devoid of large shade trees, don't despair. You can still grow these charming woodland plants in the shade of structures. Just be sure to create

appropriate soil conditions before planting by adding lots of organic matter. And be sure to keep the soil covered with mulch. Mulch is very important in the deciduous forest garden. In nature, leaves drop to the ground each fall, eventually recycling their nutrients back into the soil. Use organic mulch such as shredded leaves, shredded bark, or wood chips to create this look in your home landscape.

Because of the many small plants found in this ecosystem, you need to get up close and personal with woodland gardens. Be sure to include lots of paths made of crosscut logs or wood chips, and plant the smallest of the spring ephemerals near the paths where you can enjoy them. Benches of wrought iron, stone, or wood will offer pleasant places to enjoy the plants and provide a shady respite on hot summer days. Accent your garden with moss-covered logs, rustic stone sculptures, and planters made from cut stumps. Take advantage of the unassuming green palette and incorporate colorful sculptural pieces to add interest in summer.

Native Plants of the Deciduous Forests

A wide variety of native plants were once found in Minnesota's deciduous forests. Here are some that adapt well to cultivation.

Deciduous Trees

Acer rubrum (red maple)
Acer saccharum (sugar maple)
Betula species (birches)
Carpinus caroliniana (blue beech)
Carya ovata (shagbark hickory)
Celtis occidentalis (hackberry)
Fraxinus species (ashes)
Gleditsia triacanthos (honey locust)
Gymnocladus dioica (Kentucky coffee tree)
Juglans nigra (black walnut)
Prunus serotina (black cherry)
Quercus alba (white oak)
Quercus ellipsoidalis (northern pin oak)
Quercus rubra (red oak)
Tilia americana (basswood)

Vines

Clematis virginiana (virgin's bower)
Parthenocissus quinquefolia (woodbine)
Vitis riparia (wild grape)

Shrubs and Small Trees

Amelanchier alnifolia (Saskatoon juneberry)
Amelanchier arborea (downy serviceberry)
Amelanchier laevis (smooth juneberry)
Aronia melanocarpa (black chokeberry)
Ceanothus americanus (New Jersey tea)
Cornus species (dogwoods)
Corylus species (hazelnuts)
Diervilla lonicera (bush honeysuckle)
Dirca palustris (leatherwood)
Euonymus atropurpurea (wahoo)
Hamamelis virginiana (witch hazel)
Physocarpus opulifolius (ninebark)
Prunus species (plums and cherries)
Rhus species (sumacs)
Rosa species (wild roses)
Salix discolor (pussy willow)
Sambucus species (elders)
Spiraea alba (white meadowsweet)
Staphylea trifolia (bladdernut)
Viburnum species (viburnums)

Ferns

Adiantum pedatum (maidenhair fern)
Athyrium filix-femina (lady fern)
Matteuccia struthiopteris (ostrich fern)
Onoclea sensibilis (sensitive fern)
Osmunda cinnamomea (cinnamon fern)
Polystichum acrostichoides (Christmas fern)

Flowers and Groundcovers

Actaea species (baneberries)
Anemone canadensis (Canada anemone)
Anemone quinquefolia (wood anemone)
Anemonella thalictroides (rue anemone)
Aquilegia canadensis (Canada columbine)
Aralia racemosa (American spikenard)
Arisaema species (Jack-in-the-pulpits)
Asarum canadense (wild ginger)
Aster cordifolius (heart-leaved aster)
Caltha palustris (marsh marigold)
Campanula americana (tall bellflower)
Caulophyllum thalictroides (blue cohosh)
Claytonia virginica (Virginia spring beauty)
Dentaria laciniata (cut-leaved toothwort)
Dicentra cucullaria (Dutchman's breeches)
Dodecatheon meadia (prairie shooting star)
Erythronium americanum (yellow trout lily)
Geranium maculatum (wild geranium)
Hepatica acutiloba (sharp-lobed hepatica)
Hydrophyllum virginianum (Virginia waterleaf)
Isopyrum biternatum (false rue anemone)
Jeffersonia diphylla (twinleaf)
Lobelia cardinalis (cardinal flower)
Maianthemum canadense (Canada mayflower)
Mertensia virginica (Virginia bluebells)
Mitchella repens (partridgeberry)
Mitella diphylla (two-leaved miterwort)
Phlox divaricata (blue phlox)
Podophyllum peltatum (May apple)
Polemonium reptans (spreading Jacob's ladder)
Polygonatum biflorum (giant Solomon's seal)
Sanguinaria canadensis (bloodroot)
Smilacina racemosa (false Solomon's seal)
Thalictrum dioicum (early meadow rue)
Trillium species (trilliums)
Uvularia species (bellworts)
Viola species (violets)

Top: *Dicentra cucullaria* (Dutchman's breeches), *Sanguinaria canadensis* (bloodroot), and *Thalictrum dioicum* (early meadow rue) are all easy-to-grow, well-behaved woodland natives that adapt well to garden use.

Bottom: *Viburnum rafinesquianum* (downy arrow-wood) is a good understory shrub in woodland gardens. It is attractive all season long, starting with the reddish-tinted new leaves in spring, continuing with white flowers, and finishing the season with bluish black fruits.

Northern Coniferous Forest

The northern coniferous forest is the largest of the state's three biomes and is part of a biome stretching across North America that covers parts of the northeastern United States and Canada into Alaska. The most visible difference between coniferous and deciduous forests is the presence of evergreen conifers. Ground litter is also more acidic in conifer forests, the soil is not as rich in nutrients, and light on the forest floor is low to moderate throughout the year. The result is fewer plant species and fewer layers than in the deciduous forest. The various rock outcrops resulting from complete glaciation give this area a rugged, untamed look.

Even though most of Minnesota's original stands of white and red pine have been logged and replaced by aspen and birch forests, of the three plant communities, the northern coniferous forest has probably suffered the least at the hands of humankind. It contains fewer endangered species than deciduous forests or tall-grass prairies, and in a few places, especially along the Canadian border, forests still exit much as they did hundreds of years ago. The six coniferous-forest community types found in Minnesota are identified by the dominant species and their associated fire regimes. They are the white-pine forest, red-pine forest, jack-pine forest, black spruce-feather moss forest, spruce-balsam fir forest, and upland-cedar forest.

Plants native to this harsh environment have adapted to the cold winters and short growing season. Conifers retain their needles throughout the year so they can begin photosynthesis as soon as water is available in spring and continue well into early winter. Since they don't lose their leaves each fall, less energy is put into leaf production than in deciduous trees, so they can survive on less-fertile soils. The sparse middle layer includes seedlings and saplings of the canopy species as well as shade-tolerant shrubs such as beaked hazel, mountain maple, honeysuckle, and dogwood. The cool, shaded conditions of the forest floor make it home to many species of mosses and low-growing plants such as blue-bead lily, wintergreen, low-bush blueberry, and bunchberry that enjoy the acidic conditions brought about by the fallen pine needles.

The reduced light and acidic soil of mesic pine hardwood forests result in fewer species than are found in the other two major biomes. Ferns are an important component of this habitat.

Acer spicatum (mountain maple) is a good replacement for the overplanted nonnative amur maple (*Acer ginnala*) if you can provide it with a moist, acidic soil in partial shade.

Trientalis borealis (starflower) is a good groundcover in moist, rich, acidic soils, spreading rapidly but not too aggressively.

Distribution Limits of Native Vegetation

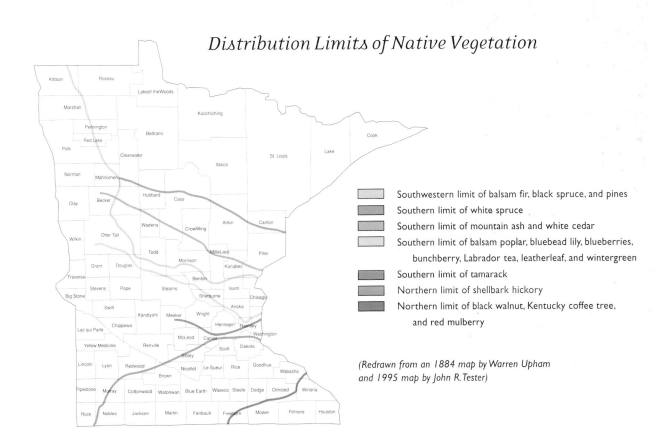

Southwestern limit of balsam fir, black spruce, and pines

Southern limit of white spruce

Southern limit of mountain ash and white cedar

Southern limit of balsam poplar, bluebead lily, blueberries, bunchberry, Labrador tea, leatherleaf, and wintergreen

Southern limit of tamarack

Northern limit of shellbark hickory

Northern limit of black walnut, Kentucky coffee tree, and red mulberry

(Redrawn from an 1884 map by Warren Upham and 1995 map by John R. Tester)

25

North Woods in the Landscape

Of the three Minnesota biomes, the northern coniferous forest is usually the most difficult for homeowners to recreate. This is because of the specific soil requirements of many of these native plants and the need for large stands of pine trees. If you don't have pines on your property but you do have shade of some sort, you can still grow many of these fascinating plants. It will just require a little more work up front. Most important is to create the necessary soil conditions. Lower the soil pH by working in lots of composted conifer needles, shredded oak leaves, or peat moss, and maintain the acidity by mulching with pine needles. The soil must also be well drained, and it can even include moderate portions of sand, gravel, and small rocks.

Plant in groups of at least five to seven plants to create the large drifts found in nature. Include lots of ferns in your plant list, and bring in fallen logs to encourage moss growth. To truly enjoy this type of garden, you must appreciate the importance of foliage and its textural qualities, since showy blooms are less prevalent than in the other biomes. The good news is that the harsh conditions of a northern coniferous habitat discourage an abundance of weeds.

If your garden is large enough to include paths, use pine needles or shredded bark. Wooden or stone benches

This successful interpretation of a northern coniferous forest habitat includes *Cornus canadensis* (bunchberry) and *Vaccinium angustifolium* (low-bush blueberry) planted in drifts for a natural look.

will blend in nicely and offer a cool respite in summer or a spot to enjoy the evergreens and showy fruits that persist into winter and keep this garden interesting year round. Limit other embellishments to fallen logs and lichen-covered rocks. A pot of colorful, shade-tolerant annuals, such as impatiens, placed on a cut stump will add summer color.

Native Plants of the Northern Coniferous Forest

Here are some plants of Minnesota's northern coniferous forests to add to your own landscape to give it that "up north" feel.

Coniferous Evergreens
Abies balsamea (balsam fir)
Picea species (spruces)
Pinus species (pines)
Taxus canadensis (Canada yew)
Thuja occidentalis (white cedar)
Tsuga canadensis (eastern hemlock)

Deciduous Trees
Acer spicatum (mountain maple)
Betula species (birches)
Larix laricina (tamarack)
Sorbus species (mountain ashes)

Shrubs
Amelanchier stolonifera
   (running serviceberry)
Arctostaphylos uva-ursi (bearberry)
Aronia melanocarpa (black chokeberry)
Chamaedaphne calyculata (leather leaf)
Comptonia peregrina (sweet fern)
Cornus species (dogwoods)
Corylus species (hazelnuts)

Dirca palustris (leatherwood)
Epigaea repens (trailing arbutus)
Ilex verticillata (winterberry)
Kalmia polifolia (bog laurel)
Ledum groenlandicum (Labrador tea)
Rosa acicularis (prickly rose)
Rubus parviflorus (thimbleberry)
Symphoricarpos albus (snowberry)
Vaccinium angustifolium
   (low-bush blueberry)

Flowers and Groundcovers
Actaea rubra (red baneberry)
Aster macrophyllus (large-leaved aster)
Calla palustris (wild calla)
Caltha palustris (marsh marigold)
Clintonia borealis (bluebead lily)
Cornus canadensis (bunchberry)
Cypripedium calceolus (yellow lady's-slipper)
Gaultheria procumbens (wintergreen)
Hepatica americana (round-lobed hepatica)
Linnaea borealis (twinflower)
Maianthemum canadense
   (Canada mayflower)

Mitchella repens (partridgeberry)
Nymphaea odorata
   (American white water lily)
Potentilla tridentata
   (three-toothed cinquefoil)
Streptopus roseus (rose twisted stalk)
Trientalis borealis (starflower)
Uvularia sessilifolia (pale bellwort)
Waldsteinia fragarioides (barren strawberry)

Ferns
Adiantum pedatum (maidenhair fern)
Athyrium filix-femina (lady fern)
Dryopteris cristata (crested fern)
Matteuccia struthiopteris (ostrich fern)
Onoclea sensibilis (sensitive fern)
Osmunda species
   (cinnamon, interrupted, and royal ferns)
Polypodium virginianum (rock-cap fern)
Polystichum braunii (Braun's holly fern)
Pteridium aquilinum (bracken)

Water Features

While Minnesota may not border an ocean, there's no denying the importance of water in the state. The state license plate may say "Land of 10,000 Lakes," but in truth, there are more than 15,000. This doesn't include the abundant wetlands, the collective term referring to wet meadows, marshes, swamps, bogs, peatlands, and prairie potholes. These areas have mostly wet soil, saturated with water either above or just below the surface, and are home to plants that have adapted to having wet feet for part or all of the growing season.

Glaciers were responsible for the formation of most of Minnesota's lakes and ponds. Ice gouged and scraped the surface of the earth, leaving depressions to be filled with melting glacial water. Wetlands occur in all three of the state's vegetative regions and can be grouped into three major types: prairie wetlands, peatlands, and forest wetlands. The distribution and look of wetlands and lakes roughly mirrors the three major plant biomes. The northern coniferous forest is home to acres of vegetation-covered peatlands and many lakes that are typically long and narrow with steep, rocky shorelines. In the deciduous forest, lakes are more gently sloping with marshy shorelines and wetlands tend toward marshes and swamps. The southeastern part of the state is home to the fewest lakes because the last glacier did not cover the area, but it has an abundance of wetlands of varying depths.

Like the other ecosystems in the state, wetlands have suffered at the hands of humans. In 1850, there were 18.6 million acres of wetlands in the state. Today only around 9 million acres exist. Most prairie wetlands and much of the deciduous forest wetlands have been drained for farmland, logging, or development. Fortunately, many of the wetlands in the northern coniferous forest remain intact.

A number of different classification systems have been developed for categorizing wetlands. One simple system used by the DNR divides them up into these cat-

Minnesota is home to numerous wetlands, ranging from bogs to the low-lying grassy areas found along ponds and lakes. *Pontederia cordata* (pickerelweed) grows successfully in the transition area along this pond.

egories: bogs, with soils made up of peat (the partially decomposed remains of plants and animals); shallow and deep marshes, the familiar open areas dominated by grasses and cattails; prairie potholes, shallow depressions formed by retreating glaciers; shrub and wooded swamps, found along the edges of lakes, rivers, and streams; seasonal basins or flats, small, isolated wetlands that contain water only seasonally along the floodplains of rivers and streams; and wet meadows, low-lying grassy areas with saturated soils often found near streams, lakes, and marshes.

Water in the Landscape

Homeowners who have a wetland, lake, pond, or stream on their property certainly have a leg up when it comes to recreating natural water features. Fortunately, it's well within the reach of most homeowners to install an artificial pond, stream, or bog. Many good references are available if your project is simple enough to fall into the do-it-yourself category. For larger projects, it is a good idea to bring in a landscape company specializing in water features. See chapter 4 for information on creating water features in the landscape.

Chapter 3

Gardening with Native Plants

Understanding Your Climate and Soil and Mastering Basic Skills

To be successful with native plants, it is essential that you understand the basics of gardening in your climate. You should have knowledge of your hardiness zone, frost dates, rainfall amounts, and existing soil conditions. You should also be aware of the basic maintenance needs of your plants and be prepared to perform them at the appropriate times.

Hardiness Zones and Frost Dates

Hardiness zones indicate the severity of winter temperatures. The lower the number, the more severe the winter climate. This book uses the most common system, the United States Department of Agriculture (USDA) Hardiness Zones, which is based on average annual minimum winter temperatures.

While it is important to know your hardiness zone, don't live and die by it. Use it as a guideline. Many factors affect a plant's ability to survive winter, such as snow cover, soil moisture, the plant's age, and winter mulching. Keep in mind that just because a plant is native in your state, it doesn't mean it will be hardy where you live. Remember, plants adapt to ecological conditions, which rarely follow state boundaries.

The growing season is the average length of time between the last killing frost in spring and the first frost in autumn. It generally increases from north to south, but it is affected by large bodies of water and other factors. Frost dates are more of a factor with tender annuals than they are with native plants. However, you should be aware of your last spring frost date, because it is used as a guide for spring planting.

Soils and Soil Preparation

Soil Texture

Good soil is the most important factor in any type of gardening, but especially with native plants. If you provide your plants with suitable soil, the rest is a piece of cake. Native plants will soon establish themselves and become almost maintenance free.

It's definitely worth taking some time up front to get to know your soil. A good place to start is to have your soil tested by a soil-testing laboratory; check with your local university extension office for labs in your area. A soil test will provide you with information on existing soil texture and fertility, along with recommendations on what to add to improve it.

Soils are typically composed of four components—sand, silt, clay, and organic matter. The proportions of these ingredients largely determine the soil texture, which in turn determines other soil properties such as fertility, porosity, and water retention. Heavier soils hold more moisture; sandy soils drain faster.

Sand and silt are the chief source of minerals required by plants, such as potassium, calcium, and phosphorous. Silt particles are smaller and yield their minerals more readily than sand, making silt soils more fertile than sandy soils. Clay particles are the finest in size, and

USDA Plant Hardiness Zones

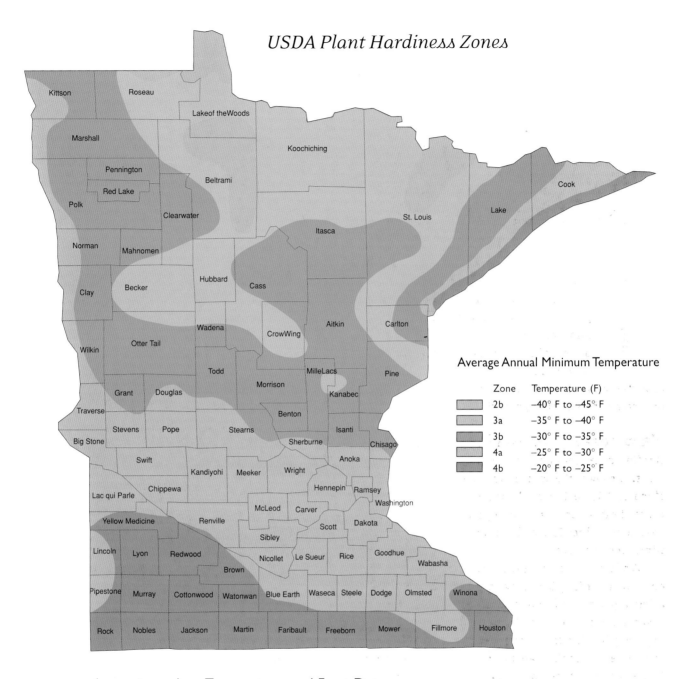

Average Annual Minimum Temperature

	Zone	Temperature (F)
	2b	−40° F to −45° F
	3a	−35° F to −40° F
	3b	−30° F to −35° F
	4a	−25° F to −30° F
	4b	−20° F to −25° F

Approximate Low Temperatures and Frost Dates

Zone	Low Temperature	Last Frost Date	First Frost Date
2	−50° F to −40° F	May 20	September 10
3	−40° F to −30° F	May 15	September 15
4	−30° F to −20° F	May 10	September 20
5	−20° F to −10° F	May 1	October 10

(Redrawn from the 2003 U.S. Department of Agriculture map)

a heavy clay soil has reduced pore spaces between the particles. These smaller spaces make it difficult for water, air, and plant roots to penetrate effectively. Clay soils generally create the greatest problem for gardeners.

Organic matter, also known as humus, is decomposing plant or animal material. It is an important component of soils and must not be overlooked in native landscapes. Organic matter determines a soil's capacity to produce nitrogen, supports the community of soil microorganisms crucial to plant life, and retains bacterial byproducts such as water and carbon dioxide. It also creates a moist, slightly acidic environment critical for the transfer of minerals from soil particles to plants.

Soil Fertility

Plants require fifteen or so nutrients for growth and survival. Carbon, hydrogen, and oxygen are involved in photosynthesis. Nitrogen, phosphorous, and potassium—each required in substantial quantities—are called macronutrients and are used to generally define soil fertility. Micronutrients necessary for healthy plants include magnesium, manganese, calcium, zinc, copper, iron, sulfur, cobalt, sodium, boron, and iodine. Since they are used in small quantities, most soils have enough for normal plant growth. Deficiencies do occur, however.

Nitrogen ensures normal vegetative growth and a healthy green color. Deficiencies result in stunted plants with a yellowish green color. Excess causes rank vegetative growth. Soil nitrogen comes mainly from decaying organic matter. Some plants also fix nitrogen on their roots. Nitrogen leaches out of soil easily, and as plants grow and remove nitrogen, more needs to be added to soil.

Phosphorus, important to flowering and fruiting, is more stable in soils than nitrogen and doesn't have to be added unless soil tests indicate a deficiency.

Potassium is essential for healthy development of roots and stems. It may need to be added to soils where plants are grown continuously.

One of the great benefits of a native-plant landscape is that if you choose your plants carefully and improve your soil regularly with organic matter, artificial fertilizers will not be necessary. In fact, they may actually be harmful, encouraging growth of exotic weeds at the expense of native plants. If your soil test indicates a need to improve soil fertility, there are several organic products you can use, including compost, rotted manure, peat moss, bone meal, blood meal, fish emulsion, soybean meal, and rock phosphate. After the initial application, yearly addition of finished compost to the surface should be enough to maintain adequate fertility.

Soil pH

Soil acidity and alkalinity are measured in terms of pH on a scale from 1 to 14, with 7 neutral. Lower than 7, the soil is increasingly acidic; higher than 7, it is increasingly alkaline. Soil pH is important because it affects the availability of nutrients necessary for plant growth. Most nutrients are most soluble at a pH between 6 and 7. That is why most plants grow best in "slightly acidic soil." Iron chlorosis, a yellowing of foliage caused by lack of available iron, can be a problem on alkaline soils.

Most native plants tolerate a range of soil pH, but some survive only within a narrow window. It is important to know which plants these are and to change the soil pH before planting them, if possible. It is possible to change soil pH after planting, but it's not as easy. As you become familiar with the native-plant communities in your area, you'll see that the plants of a community have evolved to grow best with a similar soil pH. By growing these plants together in a certain area of your garden, you'll be able to base your mulch choices and soil amendments on the pH needs of the plants. It is possible to change the soil pH around just one or two plants, however, by regularly working-in the appropriate pH-altering soil amendments.

Iron chlorosis sometimes shows up on native plants grown in soils that are too alkaline. Symptoms include a light green to yellow coloring between the leaf veins. The problem is often due to a soil pH over 6.5, which can make iron unavailable to the plants. Correct this problem by acidifying the soil to lower the pH.

Improving Soil

In general, it is best to choose plants adapted to your soil texture, pH, and moisture conditions. However, if your soil has been drastically changed by construction or other factors, you will want to do all you can to improve it before planting.

Native plants generally grow better in less-fertile soil than traditional garden plants. However, that doesn't mean they like poor soil. Most woodland plants require at least 40 percent organic matter in the soil. Most prairie plants thrive in leaner, drier soils; if too rich, plants may grow too lushly and flop over. However, some—especially those native to wetter prairies—do require richer soil.

Adding organic matter is the key to soil improvement in terms of texture, fertility, and pH. It is difficult to add too much organic matter, especially if it is partly decomposed. Organic matter increases the aeration of clay soils and improves the moisture and nutrient retention of sandy soils. It adds valuable nutrients at a slow and steady pace, and it has a buffering affect on soil pH.

In the natural world, native plants recycle themselves, creating abundant organic matter. If you can follow this principal in your own landscape, that's great. However, for city and suburban gardeners it's not always practical to allow all fallen leaves to stay on the ground and all plants to remain without cutting back.

In most situations, you will need to add organic matter on a regular basis. The best source of organic matter for gardeners is compost. Composting materials such as cow or horse manure, peat moss, grass clippings, and leaves ensures they will be in optimum condition to work into the soil. Other good sources of organic matter are chopped leaves, straw, and hay.

As with other amendments, the easiest time to add organic matter is before planting a bed. Loosen the soil with a spade or digging fork to a depth of 10 to 12 inches, more if possible. Spread a layer of compost or well-rotted manure 2 to 4 inches deep over the entire bed. Use a fork to mix it thoroughly into the soil. If your soil is very heavy (high in clay), you will want to add 2 inches of sharp builder's sand along with the compost or manure. Sand alone will only make matters worse, but when it is added to heavy clay with organic matter, it does help loosen the soil.

Organic Soil Amendments

Improve soil fertility by adding:
Compost
Well-rotted manure
Fish emulsion
Milorganite

Improve soil nitrogen content by adding:
Soybean meal
Alfalfa meal
Compost

Improve soil phosphorus content by adding:
Bone meal
Rock phosphate

Improve soil potassium content by adding:
Granite dust
Greensand
Seaweed
Wood ashes

Raise soil pH by adding:
Agricultural lime
Calcium
Wood ashes (in moderation)

Lower soil pH by adding:
Pine-needle or oak-leaf mulch
Organic matter, peat moss, or ground oak leaves
Gypsum or sulfur (in moderation)

Improve soil drainage by:
Adding compost
Adding horticultural grit
Adding coarse builder's sand with organic matter
Making raised beds

Improve water-retention by adding:
Compost
Well-rotted manure

Composting Basics

Composting is essential to successful native-plant landscaping. Compost not only adds valuable nutrients to the soil at a slow pace, it improves soil texture—both in sandy and clay soils—and it makes a great mulch. It also provides a way for you to recycle your yard waste into a useful product.

Compost is created by mixing high-carbon and high-nitrogen materials proportionately with air and moisture. High-carbon materials include straw, hay, leaves, sawdust, shredded newspaper, and pine needles. High-nitrogen materials are generally succulent green plant parts like grass clippings, weeds, perennial prunings, and vegetables. If you aren't able to provide enough of the high-nitrogen materials from your garden, you may need to add animal manure, blood meal, or cottonseed meal. Kitchen scraps such as eggshells, vegetables and fruits, and coffee grounds can also be added to the compost pile. Do not add meat scraps, bones, or grease, which attract rodents and other pests.

There is debate about whether or not you should add diseased plants or weeds with seed heads to the compost pile. Some experts feel the heat of a properly working compost pile will be enough to kill off the diseases and seeds. Others don't think it's worth the risk. To be on the safe side, you should probably avoid both.

You will need some way to contain your compost. Most people build or purchase some type of bin, but you can compost by simply piling the debris up. An ideal location is one hidden from view but close enough so that it is easy to bring stuff to it and haul the finished compost away. It is also nice to have a water source nearby so you can add water during dry spells, but it's not necessary.

Build your compost pile as materials become available, layering carbon materials alternatively with nitrogen materials. If you have an abundance of carbon materials, put some of them on the side until more nitrogen materials become available. Too many green grass clippings can mat down and prohibit the composting process. Mix them with looser materials like straw or shredded newspaper or allow them to dry in the sun before adding them to the pile. It's also a good idea to add thin layers of topsoil or finished compost to a new pile to introduce the decay organisms that create compost. Ideally, you will have several piles going at the same time so you will always have some finished compost available.

Once your pile is built, you'll need to do a little regular maintenance. Add water as needed to keep the pile moist but not soggy. Turn your compost regularly—once a week if possible—to get air into the pile. If you don't turn your pile, you'll still get compost, but it will take a lot longer. If you want to speed up the composting process, turn the pile more often, add more nitrogen-rich materials, and shred or chop the carbon materials before adding them to the pile so they break down quicker. You'll know your compost is ready for the garden when it is dark, crumbly, and most of the plant parts are decomposed.

The best way to improve soil conditions and encourage healthy plants is to apply an annual application of organic matter, such as shredded leaves and/or compost. The easiest time to apply organic matter is early spring, before the plants have fully expanded.

If possible, find a place to compost yard waste on site. Compost is an excellent soil conditioner and source of nutrients, and it provides a practical way to recycle leaves, grass clippings, and plant trimmings.

Starting a New Garden

There is a misconception among some people that native plants are tough enough that all you have to do is scatter seeds or plant an abundance of plants and they will take over an area and thrive. This misconception couldn't be farther from the truth. Native plants established themselves in an area over hundreds of years, and they can't simply be planted and expected to grow in landscapes—especially when the soil in many landscapes is far from its natural state and is covered with aggressive nonnative plants; i.e., weeds.

Proper site preparation is key to success with native plants. If you are starting a new bed, take the time to get rid of existing vegetation and improve the soil before you start putting plants in the ground. This will pay significant dividends in the end.

There are several ways to get rid of existing vegetation. A lot depends on how much time you have and how you feel about using herbicides. If your garden bed is not too large, you can dig it up manually. Just be sure to get rid of all the existing plant roots. Even tiny pieces of tough perennial-weed roots can grow into big bad weeds in no time. A major disadvantage with this method is that you lose substantial amounts of topsoil. To avoid this, if you have the time, you can simply turn the sod over and allow it to decay on site.

There are several alternatives that are easier on your back, but they take up to a year to be effective. You can smother the existing vegetation with about 6 inches of organic mulch such as straw, shredded bark, or compost. Mow closely in spring and cover with a thick layer of newspaper (ten sheets or so) and the organic mulch and let it stand all summer. Replenish the mulch in fall, and by the next spring your garden should be ready for planting. This method works best on lawn areas rather than areas with lots of deep-rooted perennial weeds.

If you don't have the luxury of a year to prepare the soil or the manual method doesn't appeal to you, you can use a nonselective glycophosphate-based herbicide such as Monsanto's Roundup, which kills tops and roots of herbaceous plants. Products based on a glycophosphate formula do not linger in the environment in a toxic form. If you follow directions exactly, aim carefully, and use only when necessary, these products should kill invading plants without causing undue harm to the environment. Once the existing vegetation is dead (usually ten days to two weeks after spraying), you can turn the soil by hand or use a mechanical tiller to turn the dead vegetation into the soil.

Avoid using a mechanical tiller without killing all existing vegetation first. While it may look like you've created a bare planting area, all you've done is ground the roots into smaller pieces, which in turn sprout into many more plants than you started with. Even after multiple tillings spaced weeks apart, you'll be haunted by these root pieces.

Whatever method you use, be careful if your garden is under large trees. Disturb soil as little as possible because digging can damage active surface roots. If you will be installing a garden right under a tree's canopy, it is best to dig individual planting holes and add organic matter to holes as needed, rather than till or dig up the entire area.

Planting

The best time to plant most plants is spring, which gives them ample time to become established before they have to endure their first winter in the ground. Summer- and fall-blooming flowers and most woody plants can be planted in spring or fall—actually, all season long if they are container grown. However, planting in the high temperatures of midsummer means you will have to be diligent about providing adequate water. You may also need to provide shelter from the sun for a few weeks. Plant or transplant early-blooming and spring-ephemeral species after they flower, which is usually late spring.

Container-grown plants can be planted at almost any time during the growing season. Be sure to dig a hole large enough to accommodate the entire root system and water well after planting. If planting in the heat of summer, provide some type of protection from the hot sun for a few days after planting.

Bare-root plants should only be planted in spring. To plant bare root, dig the hole wider and deeper than the largest roots. Make a cone of soil in the bottom of the hole, and spread the roots out over the cone. Add soil as needed to keep the plant's crown at the right level. Right after planting, water the new plants well.

Potted plants need to be carefully removed from their containers before planting. If grown in a loose soilless mix, shake off the excess and plant them like bare-root plants. If roots are circling, disentangle them to encourage outward growth. If the root ball is dense, use a sharp knife to cut through some of the roots. This may sound harsh, but your roots must be free to move into the surrounding soil.

Seeds take longer to produce a showy end product, but they are less expensive and there are often more choices available than plants. Many woodland plants have specific moisture and temperature requirements for germination and are difficult for beginners to grow from seeds. Most prairie plants, however, are relatively easy to start from seeds. If you will be seeding a large area, you may want to consider a seed mixture. However, be sure to purchase one from a reputable native-plant propagator in your area. Avoid the nationally available meadows-in-a-can.

Mulching

There are so many good things about mulching, it's hard to know where to start. Organic mulch keeps down weeds, holds water in the soil, and improves fertility. It cools the soil in woodland gardens and sets the plants off nicer than nondescript bare soil. It also replenishes the rich soil woodland plants need. Mulch around lawn trees keeps the lawn mower and weed whipper away from trunks. Keep mulch a few inches away from the base of trees to discourage mice from hiding there in winter.

There are several good organic mulches. Select one based on your plants and garden type. Shredded leaves are good in almost any situation, but they are not always available. Shredded bark or wood chips are good in shrub borders or woodland gardens, but they can be too coarse for flowerbeds. Chopped straw is good in flowerbeds but it's not as attractive as leaves. Mulch prairie gardens at planting time with a light layer of chopped, weed-free straw 1 to 2 inches deep. Once prairie gardens are established, they can be left unmulched, but it is still a good idea to top dress annually with compost. Avoid using peat moss as mulch. It tends to form a nonpermeable crust that makes it difficult for water to penetrate the soil.

The purpose of winter mulch is to offer added protection from the freezing and thawing that can occur and

Meadow-in-a-Can Seed Mixtures

Dame's rocket (*Hesperis matronalis*) is a nonnative plant often included in seed mixtures that has escaped from gardens to become a roadside weed.

Beware of meadows-in-a-can. These seed mixtures often contain many annuals that give a quick burst of color in their first year but don't reseed. Many include weedy, nonnative species that may become aggressive. If you do go this route, get the seeds from a reputable local dealer or mix your own.

Here are some plants commonly found in seed mixes that you should avoid. All are native outside the United States and can get out of control in garden settings.

baby's breath (*Gypsophila paniculata*)
bouncing bet (*Saponaria officinalis*)
chicory (*Cichorium intybus*)
cornflower (*Centaurea cyanus*)
dame's rocket (*Hesperis matronalis*)
four o'clock (*Mirabilis jalapa*)
oxeye daisy (*Leucanthemum vulgare*)
purple loosestrife (*Lythrum salicaria*)
Queen Anne's lace (*Daucus carota*)
St. John's wort (*Hypericum perforatum*)
common yarrow (*Achillea millefolium*)

Nature's mulch is the fallen layer of leaves or needles that carpet the ground each fall. Imitate nature by surrounding landscape plants with a 2- to 4-inch layer of organic mulch such as shredded bark, pine needles, or shredded leaves.

result in frost heaving of plants. It should be laid down after the ground has frozen to keep the cold in. If you put it down too early, it acts as an insulator and the ground can remain warm too long. Winter mulch should be removed in early spring just as the plants begin poking above ground.

Weed Control

A walk in the woods or through a restored prairie may lead to the mistaken idea that native landscapes don't need weeding. While it's true that a dense planting of native plants does tend to reduce weed problems, most home landscapes are a long way from their natural state, and these changes frequently invite unwelcome pests. Weeds come in from many sources, including neighbors' yards, surrounding fields, visiting birds, and even the wind. Most weeds have their origins in Europe or Asia, but not all. Certain species of *Helianthus*, *Eupatorium*, *Heliopsis*, *Viola*, and *Solidago* can become weedy in landscape situations.

Mulching is a good way to keep weeds out of your garden. However, mulch is only effective when placed on soil where the existing weeds have been removed.

Once your garden is planted and mulched, you'll still need to invest a small amount of time in weed control. Weeds are easiest to pull when they are young and the soil is moist. By taking a weekly walk through your garden after a rain, you should be able to keep weeds under control. Be sure to remove the entire plant root. Use a weeding tool to get leverage if needed.

If you have persistent perennial weeds such as thistle

The invasive common buckthorn (*Rhamnus cathartica*) was introduced to North America as an ornamental shrub, but it has now become a troublesome weed. Once established, this species aggressively invades natural areas and forms dense thickets, crowding and shading out native plants, often completely obliterating them.

or quack grass in an established garden and hand pulling has not been effective, you may want to consider spot treatments with a nonselective herbicide such as Roundup. To do this without harming yourself or nearby desirable plants, choose a calm day, protect yourself with long sleeves, safety glasses, and gloves, and carefully but thoroughly spot spray individual weeds.

In woodland gardens where woody plants such as buckthorn, poison ivy, and raspberries can become weedy, weed control will require a little more effort. You can hand pull or dig up smaller specimens, but larger plants may require the use of a nonselective herbicide. Cut the woody trunk low to the ground and use a disposable foam paintbrush to immediately apply a triclopyr-based herbicide labeled for woody plants such as poison ivy (i.e., Ortho's Brush-B-Gon) on the fresh cut. The herbicide will work its way into the trunk, destroying the plants and any plants that grow from the same root system. If you are careful and selective about using these herbicides and do not use them near water, they can effectively control weeds in the garden while causing little harm to the environment.

If you're concerned about the potential hazards of using chemical herbicides, there are safe alternative herbicides such as those made from potassium salts of fatty acids. You can also use straight vinegar, boiling water, or a butane torch.

Watering

One of the main attractions of using native plants is to reduce or even eliminate the need for supplemental watering. If you've chosen plants correctly, improved the water-holding capacity of sandy soils by adding organic matter, and have added organic mulch, established native plants rarely need supplemental watering.

Almost all plants—drought-tolerant natives included—need supplemental water while becoming established. Keep soil adequately moist until new plants have a full year of new growth on them. All trees, shrubs, and vines need to be watered regularly for at least the first full year after planting. In hot weather, they may need supplemental water once or twice a week. If the autumn is dry, continue watering until the first hard frost. Once fully established after three to four years, most native woody plants should be self-sufficient as far as watering.

Encourage deep rooting by watering less often and more deeply. The most efficient way to water is to water the soil and not the plants. Avoid overhead sprinklers. Not only are they inefficient, but the wet foliage can lead to disease problems on plants. Drip-irrigation systems and soaker hoses are good choices for effective watering.

Xeriscaping: Water-Wise Gardening

Xeriscaping promotes water conservation by using drought-tolerant, well-adapted plants within a landscape carefully designed for maximum use of rainfall runoff and minimum care. It is more common in western and southern states where rainfall is lower and water use is more regulated, but the principles behind water-wise gardening directly apply to native-plant landscaping in any area of the country.

Xeriscape landscapes are not all cactus and rock gardens. They can be green, cool landscapes full of beautiful plants maintained with water-efficient practices. Xeriscape gardening recognizes that indigenous plants are not only visually and aesthetically pleasing, they are naturally accustomed to local climates and therefore good choices for water- and waste-efficient landscapes.

Water-wise landscaping incorporates these seven basic principles, which you'll quickly realize closely match the philosophy behind native-plant landscaping:

Planning and design
Select plants based on their cultural requirements and group plants with similar water needs.

Soil analysis and improvement
Determine whether soil improvement is needed for better water absorption and improved water-holding capacity, and mix compost or peat moss into soil before planting to help retain water. Reduce water runoff by building terraces and retaining walls.

Limit lawn area
Use turf grasses as a planned element in the landscape and limit their use. Avoid impractical turf use, such as long, narrow areas, and use it only in areas where it provides functional benefits. Plant groundcovers and add hard-surface areas like patios, decks, and walkways where practical.

Plant selection
Keep your landscape more in tune with the natural environment. Select plants that have lower water requirements or those that only need water in the first year or so after planting.

Irrigate efficiently
You can save 30 to 50 percent on your water bill by installing drip or trickle irrigation systems for those areas that need watering. Install timers and water-control devices to increase their efficiency even more.

Mulch
Apply organic mulch to reduce water loss from the soil through evaporation and to increase water penetration during irrigation.

Appropriate maintenance
Properly timed pruning, weeding, pest control, and irrigation all conserve water. Raise mower blades to get a higher cut. This encourages grass roots to grow deeper, making stronger, more drought-resistant plants.

Grooming

Grooming is not about making the outdoors immaculate. It's about caring for plants and removing flowers and foliage that are past their prime in order to keep plants healthy and the garden looking pleasantly kempt. One of the advantages of native plants is that they require less grooming than nonnatives. However, some tasks will help plants grow better and keep your landscape looking more tended. These are especially important if you have a front-yard garden in the city.

Pruning

Almost every tree and shrub in your landscape will need pruning at one time or another, and the better job you do, the healthier your plants will be. Proper pruning not only encourages healthier plants, it also helps native plants, which often have a more-irregular growth habit, look neater and tidier in the landscape.

The keys to successful pruning are appropriate timing and the right tools. Most native trees and shrubs are best pruned in their dormant period, reducing the chance of infection from insects and diseases. You can also see a deciduous plant's silhouette when the leaves are off. Spring-blooming shrubs begin setting their flower buds soon after flowering and should be pruned as soon as possible after flowering. If you prune them in winter or spring, you will be cutting off their flower buds; this won't kill them, it will just mean you won't have flowers for one season. Summer-flowering shrubs are best pruned in early spring. Don't wait too long, however, or you may cut off flower buds, which start to form in spring. Evergreen conifers should be pruned in late spring and early summer just after you see new growth.

Pruning tools must be sharp and clean. The basic pruning tool is a handheld bypass pruner. Larger shrubs and small trees will require loppers and a pruning saw. If you have large trees in need of major pruning, you should have it done by a professional tree trimmer.

Pruning is somewhat of a subjective activity, but there are some basic rules to follow. Always remove dead wood and branches that rub against each other. It is usually best to try to maintain the natural shape of a tree, especially evergreens. In general, do not remove more than one-third of the branches in one pruning. However, overgrown deciduous shrubs may need renewal pruning, cutting them all the way back to the ground in early spring.

Pruning large trees is a way to alter available light. Don't be afraid to limb up large shade trees by removing some of the lower branches. This will allow more light to penetrate the garden underneath and allow you to grow a wider variety of plants.

Staking

Even with proper plant selection, taller perennials often need assistance. In a closely planted prairie garden, grasses create natural support. In the flower border, there are many systems available from nurseries using hoops and sticks of wood or metal. Use small tomato cages for bushy plants. Long-stemmed plants will need a stake for every blooming stem. Get stakes in the ground as early as possible to avoid root damage, and loosely tie plants to the stake with inconspicuous green or brown twine.

Deadheading

Removing spent flower buds stimulates prolonged and repeated blooming on many perennials. Cut back to the next set of leaves to encourage new buds to open. Think twice before deadheading plants that provide bird food and winter interest, however.

Pinching and Disbudding

Gently removing some of a plant's new growth in spring will encourage more-compact growth. This is effective on some of the taller prairie plants such as *Boltonia*, *Eupatorium*, and *Monarda* species. Pinching back will delay bloom somewhat. Do not do any pinching back after early June.

Thinning and Dividing

If you are growing perennials, ferns, and grasses in mixed borders or foundation plantings, they will benefit from dividing every three or four years. This prevents overcrowding and keeps them healthy, vigorous, and more prone to flower production.

The best time to divide most plants is early spring so they have a full growing season to recover. However, spring-flowering plants are best divided when they are dormant in fall, and vise versa, so as not to interfere with flower and seed production. To divide herbaceous plants, simply unearth the plant with a spade or trowel, wash excess soil from roots, pull or cut apart rooted sections, and replant.

Tall perennials such as *Aster novae-angliae* (New England aster) may need staking when grown in gardens. Make it easier on yourself and the plant by getting the stake in the ground early in the season while plants are still young and easy to work with.

Pest Management

Here is where native plants have a definite advantage over nonnatives. If you've spent some time preparing your site and matching plants to your site, your pest problems should be few and not serious.

The key in pest management is to really get to know your gardens and the plants in them. Keep your eyes open on daily walks. The earlier you spot problems, the easier they are to control. If you find a problem, identify it correctly (get help from a local expert, if needed), find out what is causing the problem, and decide if it is serious enough to warrant attention. Most pest problems are purely cosmetic and won't do any lasting damage to your plants. You have to decide what your level of tolerance is and determine if you want to take action or live with the problem.

Keep in mind that many problems are due to cultural conditions, not pests. Yellowing or browning foliage, stunted growth, and buds that rot before opening could be signs that you have poor soil drainage or your plants are overcrowded. Once again, making sure you have the right plant for the site and the appropriate soil conditions is the best way to avoid problems.

Native plants are by no means free of pest problems. However, many problems, such as the leaf miner damage on this columbine, are purely cosmetic and do not threaten the life of a plant. By learning to live with a few tattered leaves and less-than-perfect flowers, you'll free yourself from the need for pesticides and excessive maintenance.

Diseases and Viruses

Because native plants are adapted to a particular region, diseases and viruses are rare. If they do occur, they are rarely serious and can usually be prevented the next year by altering cultural practices, such as changing watering habits or thinning out branches. Prevention in the form of good site selection and proper planting distance is the best remedy for disease problems. If disease problems become severe, you'll need to pull up the infected plants and choose another plant for that site.

Powdery mildew, a fungal disease, is often seen on native plants, both in the wild and in gardens. To reduce chances of infection, increase air circulation by pruning out inside branches and removing some nearby plants. Cornell University has developed a baking soda-based spray for fighting black spot and powdery mildew. Mix 1 tablespoon baking soda and 1 tablespoon horticultural oil with 1 gallon of water. Spray each plant completely about once a week, starting before infections appear. It's a good idea to test the spray on a few leaves before spraying the entire plant.

Insects

Puckered foliage could indicate sucking insects, and sticky leaves often point to aphids. Do you see tiny webs on your plants? Turn over the leaves and use a magnifying glass to look for spider mites. If you catch these problems early, you can get rid of many of the pests by spraying the plants with a strong blast from your garden hose or using an insecticidal soap.

With slugs and Japanese beetles, hand picking (with gloves) and dropping them in soapy water is a good place to start. Slugs can also be killed using beer traps or diatomaceous earth placed around plants.

If any of your trees or shrubs has severely infested branches, prune them off well beyond the problem and dispose of them off site. Insects may chew and tatter leaves by the season's end. This is usually not a serious problem for most plants. Revive plants by adding compost or fish emulsion to the soil.

Natural predators keep things in check, and it is important to learn the difference between good and bad bugs. Even if you know which are the bad guys, it's still never worth it to use insecticides. Not only are they toxic to you and your garden guests, they destroy too many beneficial insects. If an insect problem becomes so bad that the health of your plant is questionable, you should consider replacing the plant with something better suited to the conditions. It's cruel to lure wildlife to your landscape and then use herbicides, insecticides, and other pesticides that can poison and destroy them.

Other Animal Pests

Insects aren't the only animal pests you'll find in your garden. Deer and rabbits are often more serious and difficult to control. There are many repellents available. However, all of them are only temporary solutions and they require a lot of time and effort to be effective, especially against deer.

Rabbits are repelled by blood meal, which can be regularly sprinkled around plants, or hot-pepper spray, which needs to be applied after every rain.

Owning a large dog can be effective in deterring deer, but the best long-term solution for a serious deer problem is to install some type of fencing, which must be at least 8 feet tall to be effective.

You can also make plant choices based on deer feeding. For starters, plant a wide variety of plants so an entire section of your landscape won't be eliminated in one meal. Although no plant can really be considered "deer proof" under all conditions, there are a few characteristics they tend to avoid. In general, deer do not like plants with thorns, aromatic plants, and plants with leathery, fuzzy, or hairy foliage. Deciduous trees are out of reach of deer browsing once they reach 6 feet in height and the bark has become corky. Young trees can be protected with a wire cage.

Planting patterns can deter deer feeding as well. Deer do not like to cross hedges or solid fences where they can't see what's on the other side. They do not like to force their way through dense shrubs or shrubs with thorns and firm branches. By massing plants, you will discourage deer from feeding in the center of the planting, where you can plant more-susceptible plants.

Deer-Resistant Native Plants

Resistance of a plant to deer feeding is affected by fluctuations in deer populations, availability of alternative food, the time of year, environmental factors, and even individual animal preference. No plant is safe under all conditions, but here are some native plants that have consistently shown some resistance to deer feeding.

Herbaceous Plants

Actaea species (baneberries)
Allium species (wild onions)
Andropogon gerardii (big bluestem)
Aquilegia canadensis (Canada columbine)
Arisaema species (Jack-in-the-pulpits)
Asclepias species (milkweeds, butterfly weed)
Asarum canadensis (wild ginger)
Aster species (asters)
Baptisia species (wild indigos)
Bouteloua curtipendula (side-oats grama)
Coreopsis species (tickseeds)
Dicentra cucullaria (Dutchman's breeches)
Echinacea species (purple coneflowers)
Eryngium yuccifolium (rattlesnake master)
Euphorbia corollata (flowering spurge)
Geranium species (wild geraniums)
Iris versicolor (northern blue flag)
Jeffersonia diphylla (twinleaf)
Liatris aspera (rough blazing star)
Lobelia species (cardinal flowers, lobelias)
Monarda species (bergamots)
Opuntia species (prickly pears)
Penstemon species (beardtongues)
Podophyllum peltatum (May apple)
Rudbeckia species (black-eyed Susans)
Ruellia humilis (wild petunia)
Sanguinaria canadense (bloodroot)
Schizachyrium scoparium (little bluestem)
Solidago species (goldenrods)
Sporobolus heterolepis (prairie dropseed)
Tradescantia species (spiderworts)
Verbena species (vervains)
Veronicastrum virginicum (Culver's root)

Woody Plants

Acer species (maples)
Amelanchier species (juneberries, serviceberries)
Betula species (birches)
Carpinus caroliniana (blue beech)
Dirca palustris (leatherwood)
Elaeagnus commutata (silverberry)
Fraxinus species (ashes)
Gleditsia triacanthos (honey locust)
Hamamelis species (witch hazels)
Ilex verticillata (winterberry)
Juniperus species (junipers)
Larix laricina (tamarack)
Picea species (spruces)
Potentilla fruticosa (shrubby cinquefoil)
Quercus species (oaks)
Tsuga canadensis (eastern hemlock)
Viburnum species (viburnums)

While no plant is guaranteed to be completely deer proof at all times, some have proven to be less palatable than others. This deer resistant planting includes *Carpinus caroliniana* (blue beech), *Asarum canadense* (wild ginger), and *Podophyllum peltatum* (May apple).

Chapter 4

Landscaping with Native Plants

Choosing Garden Styles and Incorporating Plants

Landscape design is the process of creating beautiful, useful spaces that you will enjoy being in. Your landscape should be a reflection of your lifestyle, including spaces for things you and your family enjoy doing, as well as space in which to entertain and relax. Decisions you make about your landscape should be based on practical considerations such as current conditions, future use, and maintenance issues, but don't forget to have fun and make it a place you'll truly enjoy.

A properly planted native-plant landscape will look natural on your property and with your house. Don't just carve out a geometric island bed in the lawn and fill it with plants. Aim for creating gardens that have the feel of a native habitat, keeping in mind limitations imposed by the site and proximity to neighbors. Look to nature for inspiration. Take walks in parks and nature areas. Bring a notebook and camera and record things and plants you see. Bring ideas home to help you recreate these areas—literally or by suggestion. Try to base your natural landscape on your reactions to natural beauty.

If you are starting from scratch with a new home or doing a major landscape redesign of an existing home, there are many good books available to take you through the process. Consider taking a class where a landscape designer helps you develop a plan. You may want to work with a professional landscape architect who specializes in native plants.

Assessing Your Site

Even if you're the type of gardener who doesn't like to plan everything on paper, you should have some sort of overall organized approach to your landscape. Start with a rough sketch of your property. Identify everything that is on the site, including your home and garage, existing trees and shrubs, driveways, sidewalks, outbuildings, doors, windows, utilities, faucets, downspouts, air conditioners, and planting beds. Measure everything and mark it on a map. Don't stop at the property lines. Note changes in grade and possible drainage problems. Take photographs to help you see things you may miss.

Note sunlight patterns, slopes, and compass directions. Discover and identify microclimates—those small areas where physical or biological features cause conditions to differ from the surrounding area. These can be south-facing house walls or low areas where cold air can collect.

If your lot includes a sloped shady spot where it is difficult to grow lawn grasses, stop fighting it and install a shade garden. This one mixes native plants with well-behaved nonnatives such as *Pulmonaria* species and *Athyrium nipponicum* 'Pictum' (Japanese painted fern).

Keep the architectural style of your house in mind when designing a native landscape. This prairie-style house is perfectly suited to the informal planting that surrounds it.

If you are working with a developed landscape, take note of existing plants and decide what you want to save and what you want to discard. You may be surprised at the number of native plants you already have. Note their growing conditions. Since many plants are particular about their growing requirements, you'll get clues as to what existing soil and light conditions you have.

Once you have your base drawing, it's time to determine what you want on your property. This should include practical needs as well as dreams. Chances are your list will be way too long to be practical. You'll need to prioritize and decide just what you can add at this point and what things will have to be put on hold for a while.

Choosing a Style

Once you know what you want in your landscape and have decided where to put it, it's time to think in terms of garden styles. Traditional home landscapes usually try to mimic the look of a city park—open areas with large trees and benches placed around the outside. This is rarely the most attractive or the most comfortable setting for native plants. Once again, look to nature for inspiration. Nature arranges elements according to ecological principals. Beauty in nature follows usefulness and necessity, important keys to successfully reproducing it in your own backyard.

There should be continuity between your house and landscape, but the style of your house doesn't have to dictate the style of your gardens. Most garden design books will recommend formal gardens with traditional house styles and less-formal gardens with contemporary houses, at least in the front yard. While most native plants adapt best to naturalistic designs, they can be used in formal settings. You'll just have to put a little more thought into plant selection and planting design. See the sidebar (page 42) for ideas on how to make native plants work in more traditional settings.

When planning your landscape, decide what kind of habitats you want to include. Determine what indigenous plant community makes ecological sense in each spot. For best results, try to mimic a natural habitat compatible with your conditions. If your site is wooded, marshy, or otherwise undeveloped, the simplest and soundest course is simply to build on the natural habitat. Rather than fill in the wet spot, stock it with native wetland species and create a marsh, pond, or bog. If you are tired of fighting with grass under large shade trees, remove the grass and begin creating a deciduous woodland garden. If you have a large open lawn area in full sun that is just a drain on your lawn mower and your time, consider a prairie garden. There are native plants to fit most popular garden styles and landscape situations.

There are many things suburban gardeners can do to help make their native landscape more acceptable to their inexperienced neighbors, such as maintaining some turf areas, using mulch, and planting clumping plants rather than spreaders.

A Well-Tended Native Garden

Even with all the benefits of using native plants, there are still people who have a hard time appreciating them in the landscape. If you live in an urban area and are concerned about what the neighbors will think, here are some things you can do to make your native landscape look more tended and less wild.

❖ Maintain at least a small area of turf grass
❖ Use a buffer of grass or mulch between planted areas and sidewalks and streets so plants don't flop over onto the paved areas, making visitors uncomfortable
❖ Cut back plants in fall
❖ Avoid planting excessively tall plants
❖ Prune trees and shrubs so they grow more open, like nursery-grown trees
❖ Add embellishments such as sculpture and benches to make it look more like a garden and less like a field of weeds

❖ Install some straight edges, either around garden beds or in the form of paths and fences
❖ Plant natives in a more traditional way with even spacing between plants, and mulch with cocoa-bean hulls or shredded bark
❖ Limit your number of species
❖ Plant in clumps as is more typical of nonnative landscapes
❖ Use natives that are "more controllable" in their growth habit; these tend to be the clump-forming plants rather than spreaders

Placing Plants

When it comes to choosing what to grow and where to grow it, most books will overwhelm you with design lingo about form, texture, color, and so on. While you should keep these things in mind, landscaping with native plants frees you up from much of this. Nature has already made these decisions for you. If you follow her lead, you'll end up with a landscape full of interesting textures, colors, and plant forms without much effort. Plants found growing together in the wild will usually combine nicely in a garden setting.

It should be clear by now that landscaping with native plants focuses more on your existing site conditions and matching plants to them rather than selecting plants first—a holistic approach, rather than thinking in terms of individual plants. By grouping plants with similar soil, water, and light requirements, you'll end up with combinations that work well together naturally.

That said, even with all the information geared toward choosing the right plant for the right place, don't be afraid to try a few things that intrigue you, even if odds are against survival. Make a point of growing something new each year. Even if it doesn't survive, you'll gain confidence and skill from the experience.

There is a misconception that natural landscapes are chaotic and lacking in perceptible patterns. In relatively undisturbed, naturally evolving landscapes, patterns are ever-present—not in the form of orchardlike grids of trees, but in the subtle arrangements of plants. Nature tends to mass similar forms together and accent them with contrasting forms. Take a cue and use contrast sparingly to give your garden a natural look. Texture comes from foliage as well as bark and stems, stone walls, and pathways. Avoid the neat, orderly rows of plants common in formal gardens; nature's pattern is randomness. Cluster plants in odd-numbered groupings rather than scattering them singly here and there.

Plant in drifts of color rather than in straight rows. Drifts can consist of several of the same plants or, for the plant collectors among us, single plants of similar colors from several different species. Place an emphasis on choosing plants with interesting foliage shapes and textures, since foliage is usually decorative much longer than flowers are. Incorporate a variety of foliage colors into borders, including traditional greens in all shades, silvers, blues, chartreuse, and the occasional variegated form. Plants with light-colored flowers and silver-leaved foliage glow in the moonlight and are ideal in a garden next to a deck or patio that will be used at night.

Make sure your landscape has a wide variety of species—both within it and different from all the other trees and shrubs on the street. Monocultures can lead to big problems in the landscape, as we saw with the American elm. Go for diversity, especially with woody plants.

Choose plants with varying bloom times, both concurrent and consecutive, to ensure that there is something happening in your landscape all year long. Evergreens are essential to winter landscapes, but many other woody ornaments offer interesting bark and colorful berries or rose hips that persist through winter. Plants with showy persistent fruits include Jack-in-the-pulpits, bunchberry, wintergreen, winterberry, partridgeberry, false Solomon's seal, and highbush cranberry. Allow goldenrods, milkweeds, rattlesnake master, Joe-pye weeds, and grasses to remain through the winter so their dried flower heads can provide visual interest, as well as food for birds. Include fall-blooming perennials that survive several degrees of frost to prolong the growing season.

For a sense of unity, repeat a few specific plant groupings or color schemes at intervals throughout the landscape. Include some areas of visual calm where the eye can rest momentarily from stimulation. Green lawns, a small grouping of silver-leaved plants, or a simple green deciduous or evergreen shrub all create spots of calm.

Choose plants that appeal to all the senses: plants in a variety of colors, shapes, and textures to see; fragrant blossoms and foliage to smell; berries to eat; soft, hairy foliage, smooth bark, and silky seed heads to touch; leaves that whisper and rustle in the wind to hear. Plant fragrant species close to seating areas or under windows where they can be fully appreciated.

If space allows, let woodland flowers and groundcovers spread and seed themselves to form natural drifts like those found in nature. An occasional clump of snow-white *Trillium grandiflorum* (large-flowered trillium) provides accent and interest in this planting, which includes *Phlox divaricata* (blue phlox), *Mitella diphylla* (two-leaved miterwort), and ferns.

Allowing native grasses such as *Schizachyrium scoparium* (little bluestem) and the seed heads of *Rudbeckia hirta* (black-eyed Susan) to remain over winter not only provides birds with food and shelter, it also adds interest to the landscape.

Creating Mixed Borders

The mixed border is a plant-lover's garden. This combination of woody and herbaceous plants can include small trees, shrubs, flowers, grasses, groundcovers, ferns, and even evergreens. It is a place to experiment with colors and textures, combining plants that not only grow well together, but also complement each other.

A well-planned mixed border will have color and interest year round. Mix heights, shapes, and textures to give depth and dimension, with taller, spikier plants generally located toward the back, and shorter, clumpier plants closer to the front of the garden or along paths. For contrast, plant one or two taller see-through plants toward the front of the border.

Many native plants combine well with nonnatives in mixed-border situations. Most are natives of prairie and savanna ecosystems and require full to part sun. In general, plant in groups of three to five; however, don't be afraid to use single plants here or there for accent and interest. If you want to create more of a prairie-garden look, avoid planting nonnatives and go heavier on the grasses, keeping in mind that natural prairies consist of roughly 80 percent grasses and 20 percent flowers.

Geum triflorum (prairie smoke) is an excellent plant for mixed borders. Showy rosy pink flowers appear early in spring, the foliage is attractive all growing season, and the heads of feathery plumes add interest in late summer.

Rudbeckia hirta (black-eyed Susan) may well be the most popular native plant found in home landscapes, and rightly so. It is easy to care for and brings long lasting color to perennial beds and mixed borders.

Native Plants for Mixed Borders

Shrubs and Small Trees

Amelanchier × grandiflora (apple serviceberry)
Amorpha canescens (leadplant)
Ceanothus americanus (New Jersey tea)
Cephalanthus occidentalis (buttonbush)
Cornus alternifolia (pagoda dogwood)
Dirca palustris (leatherwood)
Physocarpus opulifolius cultivars (ninebark)
Picea cultivars (dwarf spruces)
Pinus cultivars (dwarf pines)
Potentilla fruticosa cultivars (shrubby cinquefoil)
Prunus nigra 'Princess Kay' (Canada plum)
Spiraea alba (white meadowsweet)

Flowers

Agastache foeniculum (blue giant hyssop)
Allium cernuum (nodding wild onion)
Aquilegia canadensis (Canada columbine)
Asclepias tuberosa (butterfly weed)
Aster species (asters)
Baptisia species (wild indigos)
Boltonia asteroides (boltonia)
Campanula species (harebells, bellflowers)
Dalea purpurea (purple prairie clover)
Echinacea species (purple coneflowers)
Eryngium yuccifolium (rattlesnake master)
Eupatorium species (Joe-pye weeds)
Gaillardia species (blanket flowers)
Gentiana species (gentians)
Geum species (prairie smokes)
Helenium autumnale (autumn sneezeweed)
Heliopsis helianthoides (oxeye)
Liatris species (blazing stars)
Lobelia species (cardinal flower, lobelias)
Lupinus perennis (wild lupine)
Mertensia virginica (Virginia bluebells)
Monarda species (wild bergamot, bee balms)
Penstemon species (beardtongues)
Phlox species (phloxes)
Physostegia virginiana 'Miss Manners' (obedient plant)
Polemonium reptans (spreading Jacob's ladder)
Pulsatilla patens (pasque flower)
Ratibida species (coneflowers)
Rudbeckia species (black-eyed Susans)
Ruellia humilis (wild petunia)
Senecio aureus (golden ragwort)
Sisyrinchium species (blue-eyed grasses)
Solidago rigida, S. speciosa (goldenrods)
Tradescantia species (spiderworts)
Verbena species (vervain)
Veronicastrum virginicum (Culver's root)
Zizia aptera (heart-leaved alexanders)

Ferns and Grasses

Matteuccia struthiopteris (ostrich fern)
Panicum virgatum (switch grass)
Schizachyrium scoparium (little bluestem)
Sorghastrum nutans (Indian grass)
Sporobolus heterolepis (prairie dropseed)

Amelanchier 'Princess Diana' is a good small tree for mixed borders, offering year-round interest—delicate white flowers in spring, clean foliage and showy berries in summer, good fall leaf color, and smooth gray bark for winter interest.

Shade Gardening

Many gardeners consider shade to be a liability, a spot where they can't grow grass or colorful flowers. However, all landscapes should have a cool, shady retreat from the summer heat, and some of the most interesting plants grow in these areas protected from the hot sun. With a little effort, you can eliminate the struggle with lawn grasses and fill the area with a beautiful tapestry of colors and textures of native plants that thrive in shade. And, a shade garden is a joy to tend. It is usually less weedy than sunny gardens and much more comfortable to work in, especially on a hot summer day.

The key to successful shade gardening is understanding the different degrees of shade, which can range from light to heavy shade and are different at different times of the growing season. "Light shade" areas receive bright to full sun for all but a few hours each day. Areas with bright light or sun for about half the day are called "partial shade." Most shade plants will do fine in either of these sites, especially if the sun is morning sun. "Full shade" areas are shaded for most of the day, and "dense shade" is for only the most shade-tolerant plants. Don't be afraid to limb up some taller trees to provide the partial shade that so many woodland wildflowers thrive under.

When it comes to plant selection, simplicity, rather than ostentation, is the key. Textures and shades of green play an important role. If possible, allow flowers and groundcovers to spread and seed themselves and to form natural drifts. You want plants that are adapted, attractive, and long-lived, and plants that increase, but not so rapidly that they crowd out other desirable plants. Some plants that have many desirable characteristics—such as wild ginger, baneberries, violets, and May apple—may be too much for smaller gardens.

Most woodland plants require a slightly acidic, fertile soil. Add 2 to 4 inches of organic matter when building the garden or at planting time. Watering is important, especially where large tree roots compete for available moisture. Mulch with 2 to 4 inches of shredded leaves, bark, or pine needles to conserve moisture and replenish nutrients. If your garden is large enough, include curved paths covered with wood chips or shredded bark. Create points of interest along the way—ornaments such as wagon wheels or hollow stumps filled with showier flowers; larger patches of flowers; or trees and shrubs with unique growth patterns or bark that are enhanced by pruning. Allow some branches and logs to remain after falling. Consider adding a trickling stream lined with marsh marigolds to add sound.

Shade gardens are at their best in spring, when the delicate woodland flowers and bright green foliage are set off by the rich color of the bare ground. *Mitella diphylla* (two-leaved miterwort) and *Podophyllum peltatum* (May apple) are two good choices for light to full shade.

Uvularia grandiflora (large-flowered bellwort) has long-blooming, lemon yellow flowers that provide a striking contrast to the more common white, pink, and purple flowers of spring and attractive sea green foliage in summer. It likes spring sun and summer shade, conditions found under deciduous trees.

Native Plants for Shade Gardens

Woody Plants
Abies balsamea (balsam fir)
Amelanchier species (serviceberries, juneberries)
Cornus species (dogwoods)
Dirca palustris (leatherwood)
Hamamelis virginiana (witch hazel)
Ostrya virginiana (ironwood)
Tsuga canadensis (eastern hemlock)
Viburnum species (viburnums)

Ferns
Adiantum pedatum (maidenhair fern)
Athyrium filix-femina (lady fern)
Dryopteris marginalis (marginal shield fern)
Matteuccia struthiopteris (ostrich fern)
Osmunda species (ferns)
Polystichum acrostichoides (Christmas fern)

Flowers and Groundcovers
Actaea species (baneberries)
Allium tricoccum (wild leek)
Anemone species (anemones)
Anemonella thalictroides (rue anemone)
Aquilegia canadensis (Canada columbine)
Arisaema triphyllum (Jack-in-the-pulpit)
Asarum canadense (wild ginger)
Aster macrophyllus (large-leaved aster)

Carex species (woodland sedges)
Caulophyllum thalictroides (blue cohosh)
Claytonia virginica (Virginia spring beauty)
Dentaria laciniata (cut-leaved toothwort)
Dicentra species (Dutchman's breeches, bleeding hearts)
Erythronium species (trout lilies)
Hepatica species (hepaticas)
Isopyrum biternatum (false rue anemone)
Jeffersonia diphylla (twinleaf)
Lobelia cardinalis (cardinal flower)
Maianthemum canadense (Canada mayflower)
Mertensia virginica (Virginia bluebells)
Mitchella repens (partridgeberry)
Mitella diphylla (two-leaved miterwort)
Phlox divaricata (blue phlox)
Podophyllum peltatum (May apple)
Polemonium reptans (spreading Jacob's ladder)
Polygonatum species (Solomon's seals)
Sanguinaria canadensis (bloodroot)
Smilacina species (false Solomon's seals)
Solidago flexicaulis (zigzag goldenrod)
Thalictrum dioicum (early meadow rue)
Trillium species (trilliums)
Uvularia species (bellworts)
Viola species (violets)

Nonnative Plants for Shade
Here are some well-behaved nonnative shade plants that combine nicely with the subtle beauty of native woodland plants.

Alchemilla mollis (lady's mantle)
Asarum europaeum (European ginger)
Astilbe species and hybrids (astilbes)
Athyrium nipponicum 'Pictum' (Japanese painted fern)
Brunnera macrophylla (brunnera)
Dicentra species (bleeding hearts)
Epimedium species and hybrids (barrenworts)
Galium odoratum (sweet woodruff)
Helleborus niger (Christmas rose)
Helleborus orientalis (Lenten rose)
Heuchera species and hybrids (heucheras)
×*Heucherella* hybrids (heucherellas)
Hosta species and hybrids (hostas)
Lamium maculatum (lamium)
Ligularia species (ligularias)
Pachysandra terminalis (Japanese spurge)
Phlox stolonifera (creeping phlox)
Primula species (primroses)
Pulmonaria species (lungworts)

Attracting Butterflies

Butterflies add a wonderful dimension to the landscape, and they are one of the best side benefits of having a native-plant landscape. Your efforts at butterfly gardening can be as simple as incorporating a few nectar plants into a flower garden or as elaborate as creating an area entirely devoted to these fascinating creatures.

To become a successful butterfly gardener, you should start by learning which butterfly species are native to your area and what plants they like. Most butterflies are vagabonds on their way somewhere else when you see them in your garden. You can think of your garden as a rest stop along the way, a place where they can linger for a while to enjoy food, water, and shelter. If you want to provide a complete butterfly habitat, you will have to include the proper host plants for butterflies to lay eggs, keeping in mind that these are not always the showiest plants and that they become even less attractive when they have been eaten by the newly hatched caterpillars. They are important, however, as without certain host plants, butterflies will not lay eggs.

When it comes to nectar sources, these insects are drawn to big, bold splashes of color and concentrations of fragrance. Butterflies like two kinds of flowers: clusters of nectar-filled tubular blossoms that they can probe in sequence, and large, rather flat blossoms that provide them with landing pads. Purple, red, orange, and yellow attract the most butterfly species; blue and white flowers are least popular. However, color preferences can vary from species to species, so you should plant the entire color palette. Especially valuable for butterflies are early blooming flowers that are open when the first hatches emerge or the first returning migrants arrive, and late-blooming plants that are still in bloom after the first fall frost. Early violets are occasionally weighted down with butterflies, as are late-blooming asters and Joe-pye weeds.

In addition to host plants, butterflies need mud puddles or other wet areas, and shelter—shrubs, trees, and bushy flowers where they can hide from birds, find shade at midday, and rest at night. They also look for basking stones, where they can build up enough body heat to fly, and windbreaks to temper the wind.

Agastache foeniculum (blue giant hyssop), a good source of nectar, becomes even more attractive to butterflies when a shallow water source such as a birdbath is nearby.

With its intense orange-red flowers, *Asclepias tuberosa* (butterfly weed) is one of the showiest and easily recognized of native flowers. Butterflies agree, and the flowers are covered with them in June and July.

Native Plants to Attract Butterflies

Here are some native plants that are nectar sources as well as food for caterpillars. Many of these plants can be incorporated into large sunny gardens where you can plant drifts. If you want to include several of the less-showy nectar sources, you may want to set aside a special area of your landscape specifically for a butterfly habitat. "N" indicates it is a good adult nectar source; "L" indicates a larval food source.

Trees and Shrubs
Amorpha species (leadplant, false indigo) N
Betula species (birches) L
Ceanothus americanus (New Jersey tea) N
Celtis occidentalis (hackberry) N
Cephalanthus occidentalis (buttonbush) N
Cornus species (dogwoods) N
Ledum groenlandicum (Labrador tea) N
Populus species (poplars) L
Prunus species (plums) N
Quercus species (oaks) L
Rhus typhina (staghorn sumac) N
Rosa species (wild roses) N
Salix species (willows) L
Sambucus species (elders) N
Spiraea alba (white meadowsweet) N
Symphoricarpos species
   (snowberry, wolfberry) L
Viburnum species (viburnums) N

Herbaceous Plants
Agastache foeniculum (blue giant hyssop) N
Anaphalis margaritacea
   (pearly everlasting) N
Antennaria species (pussytoes) N
Asclepias species
   (milkweeds, butterfly weed) N, L
Aster species (asters) N, L
Baptisia species (wild indigos) N
Carex species (sedges) L
Chelone glabra (white turtlehead) N
Coreopsis species (coreopsis, tickseeds) N
Dalea purpurea (purple prairie clover) N
Echinacea species (purple coneflowers) N
Eupatorium species (Joe-pye weeds) N
Fragaria virginiana (wild strawberry) N
Gaillardia species (blanket flowers) N
Helianthus species (sunflowers) N
Heliopsis helianthoides (oxeye) N

Liatris species (blazing stars) N
Lupinus perennis (wild lupine) L, N
Monarda species
   (wild bergamot, bee balms) N
Penstemon species (beardtongues) L
Phlox species (phloxes) N
Pycnanthemum species (mountain mints) N
Ratibida species (coneflowers) N
Rudbeckia species (black-eyed Susans) N, L
Schizachyrium scoparium (little bluestem) L
Silphium species
   (compass plant, cup plant) N
Solidago species (goldenrods) N
Verbena species (vervains)
Viola species (violets) L
Zizia species (alexanders) L

Creating a Hummingbird Habitat

Hummingbirds bring movement to a garden, mesmerizing you as they zoom from one plant to the next. There's nothing more enjoyable than a visit from these tiny, wing-flapping guests.

Because of its high rate of metabolism, a hummingbird needs to eat more than one-half its weight in food daily. A hummingbird habitat should include several types of flowers—herbaceous and woody plants of varied heights and bloom dates. Your hummingbird habitat can also include several properly maintained feeders. The most effective habitats also attract and nurture tiny insects and spiders that hummingbirds ingest to meet their protein requirements.

Since hummingbirds, like most birds, have virtually no sense of smell, the flowers that attract them tend to have little or no fragrance, instead direct resources toward high visibility and nectar production. Note also that cultivated hybrids often make much less nectar than wild strains. Most hummingbird-attracting flowers are tubular in shape and many are red, though certainly not all. A successful hummingbird garden provides nectar sources from May through the first frost. There is a great temptation to plant acres of wild bergamot or cardinal flower, two of the hummingbird's favorite nectar sources. However, with each of these flowers, nectar is available for just a brief period in a hummingbird's life.

Your garden should also have space for hummingbirds to nest and locations where they can roost and find shelter from the elements. Have fresh water available for drinking as well as for bathing. Include shady spots where hummingbirds can perch as well as build their nests. Willows provide nesting materials that hummingbirds will use. They also look for bits of leaves, spider webs, moss, and lichens to build their tiny nests.

Hummingbirds are attracted to red flowers, so it's no surprise that showy *Lobelia cardinalis* (cardinal flower) is at the top of every hummingbird plant list.

The nodding, upside-down, red and yellow flowers of *Aquilegia canadensis* (Canada columbine) are a favorite of hummingbirds.

Native Plants to Attract Hummingbirds

Here are some native plants that are important nectar sources for hummingbirds. You'll notice that many of them are also attractive to butterflies.

Woody Plants
Ceanothus americanus (New Jersey tea)
Parthenocissus quinquefolia (woodbine)
Symphoricarpos albus (snowberry)

Herbaceous Plants
Agastache foeniculum (blue giant hyssop)
Aquilegia canadensis (Canada columbine)
Asclepias tuberosa (butterfly weed)
Chelone species (turtleheads)
Mertensia virginica (Virginia bluebells)
Epilobium angustifolium (fireweed)
Liatris species (blazing stars)
Lilium species (lilies)
Lobelia cardinalis (cardinal flower)
Monarda species (wild bergamot, bee balms)
Penstemon species (beardtongues)
Phlox species (phloxes)
Physostegia virginiana (obedient plant)
Silphium perfoliatum (cup plant)

Gardening among Rocks

True rock gardens are made up of low-growing plants that inhabit rocky areas in high elevations. However, rocks themselves are an important feature of many natural habitats, including outcrops, woodlands, and along streams, and they can be a real asset in naturalistic landscape design. You can use a few here and there for accent, or you can create a "rock garden" where suitable plants are planted amongst a rock-filled hillside.

If you are starting from scratch, there are a few things you can do to make your garden look more natural. Select an indigenous rock and stick with it throughout the entire garden, incorporating a range of sizes. Bury the rocks at least halfway to make them look like they were positioned by nature and have been there through time. To be convincing, plant placement calls for an appreciation of how they grow naturally on rocky sites. Usually irregular drifts of low plants carpet the soil surface or occupy the spaces between rocks, running along narrow crevices or congregating in areas at the bases of miniature cliffs. A sunny rock garden is one of the few places where gravel mulch is appropriate.

Antennaria plantaginifolia (pussytoes) and *Pulsatilla patens* (pasque flower) are low-growing, drought-tolerant flowers that adapt well to sunny rocky areas.

Native Plants for Rock Gardens

Rock-garden plants tend to be low-growing and tolerant of hot, dry conditions, but there are also plants that are suitable for shadier sites among rocks edging pathways and streams. Here are some native plants to use among rocks.

Plants for Sunny Spots

Allium cernuum (nodding wild onion)
Allium stellatum (prairie wild onion)
Anaphalis margaritacea (pearly everlasting)
Antennaria species (pussytoes)
Aquilegia canadensis (Canada columbine)
Asclepias tuberosa (butterfly weed)
Aster sericeus (silky aster)
Bouteloua curtipendula (side-oats grama)
Campanula rotundifolia (harebell)
Coreopsis species (coreopsis, tickseeds)
Dalea purpurea (purple prairie clover)
Dodecatheon meadia (prairie shooting star)
Echinacea species (purple coneflowers)
Eryngium yuccifolium (rattlesnake master)
Euphorbia corollata (flowering spurge)
Gaillardia species (blanket flowers)
Geum species (prairie smokes)
Liatris punctata (dotted blazing star)
Gaura species (gauras)
Opuntia species (prickly pears)
Penstemon species (beardtongues)
Phlox pilosa (prairie phlox)
Physostegia virginiana (obedient plant)
Potentilla fruticosa (shrubby cinquefoil)
Potentilla tridentata (three-toothed cinquefoil)

Pulsatilla patens (pasque flower)
Ratibida species (coneflowers)
Ruellia humilis (wild petunia)
Sisyrinchium species (blue-eyed grasses)
Solidago nemoralis (gray goldenrod)
Tradescantia species (spiderworts)
Viola pedata (bird's-foot violet)
Waldsteinia fragarioides (barren strawberry)

Plants for Shady Spots

Adiantum pedatum (maidenhair fern)
Anemonella thalictroides (rue anemone)
Aquilegia canadensis (Canada columbine)
Asarum canadense (wild ginger)
Asplenium platyneuron (ebony spleenwort)
Carex pensylvanica (Pennsylvania sedge)
Cystopteris bulbifera (bulblet fern)
Dodecatheon meadia (prairie shooting star)
Dryopteris marginalis (marginal shield fern)
Gaultheria procumbens (wintergreen)
Hepatica species (hepaticas)
Mitchella repens (partridgeberry)
Phlox divaricata (blue phlox)
Polypodium virginianum (rock-cap fern)
Thalictrum dioicum (early meadow rue)
Uvularia species (bellworts)
Viola sororia (common blue violet)

Rocks play an important role in native landscape design. Use them in shade gardens to add texture and interest.

Water Gardens and Bog Gardens

Water, whether moving or still, is a great addition to the landscape. A water feature offers boundless possibilities for creative design as well as a place to grow many interesting native species that require the conditions found in these ecosystems.

If you have a natural wet area on your property, instead of altering the habitat by drainage or fill, work within the parameters of the existing environment to create a pond, bog garden, or marshy habitat. If you don't have a natural water feature on your property, it is fairly easy to create a pond or bog garden.

To make your water feature look natural, think about how pools of water occur in nature. The pond will look most natural in the lowest place in the yard. However, avoid areas where runoff collects. Runoff can cloud the water or fill the pond with leaves, mulch, and other debris after heavy rain. If your yard is flat, you can site the pond just about anywhere, but it should get at least five hours of sunlight a day for good bloom on water lilies and other flowering aquatics.

Also, consider the size and proportion in relationship to your yard. A 5x8-foot pond is a generous size even in a large backyard and can accommodate many plants. A 3x5-foot pond is appropriate for a small space. Place it as nature would—with no corners, no symmetry, nothing perfectly round. Avoid rigid pond liners that come in precast shapes.

Bog gardens are areas of permanently moist but not waterlogged soil. A bog garden can be created by excavating to a depth of 15 to 18 inches and covering the bottom with a tarp or pool lining material to retard drainage. Fill the depression with the removed soil and amend generously with peat moss or other organic matter. Make slits in the liner to allow seepage during wet weather and plan to add supplemental water during periods of low rainfall. Fill your garden with moist-soil and shallow-water plants.

Planting techniques for aquatic and wetland species are basically the same as those used with other plants. Do not allow the plants to dry out while planting, and plant them at the proper level—not only to the soil but also to the depth of the water over the crown.

Aquatics can be planted directly in a thick layer of soil at the bottom of a pond, but they are best planted in containers because many are aggressive and quickly become unmanageable if not restricted. Planting in containers also makes it easier to remove them or shift their position. You can use specially designed plastic water-lily baskets or used 3-gallon nursery containers with the sides perforated.

Aquatics should be planted in good, clean garden soil, free from organic matter and low in nutrients, both of which encourage algae. Do not fertilize garden pools.

To create the feel of a natural water feature, place plants in concentric rings radiating from your water feature, with the plants requiring the most soil moisture closest to the pond or stream. Emergent aquatics inhabit the margins of ponds and slow-moving streams. They are used to soften the edge of ponds and streams with contrasting growth habits and textures. Many native emergent aquatics provide additional summer color and homes for many forms of wildlife. They should be grown in shallow water or in the mucky soil bordering water features.

Many native plants adapt well to the soggy soil conditions found along streams and ponds. This planting features *Lobelia cardinalis* (cardinal flower), *Rudbeckia triloba* (three-leaved coneflower), *Eupatorium purpureum* (Joe-pye weed), *Chelone glabra* (white turtlehead), and *Vernonia fasciculata* (bunched ironweed).

Rainwater Gardens

A rainwater garden is a great way to create a useful wetland that helps control runoff from impervious surfaces and filter out pollutants before they reach streams and lakes.

Rainwater gardens are shallow—usually less than a foot deep—and range in size from gardens as small as the area under your downspout to large gardens covering several city blocks. By slowing down stormwater runoff, rainwater gardens collect water and allow it to slowly seep into the soil.

Planted with appropriate native species, rainwater gardens become attractive additions to the landscape and a haven for butterflies and birds. Any of the plants listed in the sidebar as "moist-soil" plants are suitable for use in a rainwater garden.

This successful wetland landscape includes *Nymphaea odorata* (American white water lily) and moisture-loving edging plants, which make the transition to the neighboring prairie area.

Native Plants for Water Gardens

Pond Plants

Nymphaea odorata
  (American white water lily)

Shallow-Water Plants
For planting in water levels less than 14 inches:

Calla palustris (wild calla)
Carex stricta (tussock sedge)
Cephalanthus occidentalis (buttonbush)
Glyceria species (manna grasses)
Pontederia cordata (pickerelweed)
Sagittaria latifolia (broad-leaved arrowhead)

Plants for Moist-Soil Areas
Grasses

Calamagrostis canadensis (blue joint)
Deschampsia caespitosa (tufted hair grass)
Panicum virgatum (switch grass)
Spartina pectinata (prairie cordgrass)

Ferns

Most ferns will do well in moist soil, but these are especially well suited to planting along ponds and streams and in bogs.
Athyrium filix-femina (lady fern)
Cystopteris bulbifera (bulblet fern)
Matteuccia struthiopteris (ostrich fern)
Onoclea sensibilis (sensitive fern)
Osmunda species (ferns)

Woody Plants

Acer rubrum (red maple)
Acer saccharinum (silver maple)
Alnus incana (speckled alder)
Amorpha fruticosa (false indigo)
Aronia melanocarpa (black chokeberry)
Betula alleghaniensis (yellow birch)
Betula nigra (river birch)
Cephalanthus occidentalis (buttonbush)
Cornus racemosa (gray dogwood)
Cornus stolonifera (red-osier dogwood)
Fraxinus nigra (black ash)
Fraxinus pennsylvanica (green ash)
Ilex verticillata (winterberry)
Larix laricina (tamarack)
Ledum groenlandicum (Labrador tea)
Picea mariana (black spruce)
Quercus bicolor (swamp white oak)
Salix species (willows)
Sambucus canadensis (common elder)
Spiraea alba (white meadowsweet)
Thuja occidentalis (white cedar)
Viburnum trilobum (highbush cranberry)

Flowers

Anemone canadensis (Canada anemone)
Arisaema triphyllum (Jack-in-the-pulpit)
Asclepias incarnata (swamp milkweed)
Aster novae-angliae (New England aster)
Caltha palustris (marsh marigold)

Campanula americana (tall bellflower)
Chelone species (turtleheads)
Cornus canadensis (bunchberry)
Epilobium angustifolium (fireweed)
Equisetum hyemale (tall scouring rush)
Eupatorium species (Joe-pye weeds)
Gentiana andrewsii (bottle gentian)
Helenium autumnale (autumn sneezeweed)
Iris species (blue flags)
Liatris ligulistylis
    (northern plains blazing-star)
Liatris pycnostachya (great blazing star)
Lilium michiganense (Michigan lily)
Lobelia cardinalis (cardinal flower)
Lobelia siphilitica (great lobelia)
Lysimachia ciliata (fringed loosestrife)
Mertensia virginica (Virginia bluebells)
Physostegia virginiana (obedient plant)
Pycnanthemum virginianum
    (Virginia mountain mint)
Rudbeckia triloba (three-leaved coneflower)
Senecio aureus (golden ragwort)
Silphium perfoliatum (cup plant)
Thalictrum dasycarpum (tall meadow rue)
Verbena hastata (blue vervain)
Vernonia fasciculata (bunched ironweed)
Veronicastrum virginicum (Culver's root)
Viola sororia (common blue violet)

Hardscapes and Accents

The hardscape is anything in your landscape that isn't living. It includes paths, fences, decks, patios, driveways, and all structures such as arbors, compost bins, and stairways. These structures are important parts of the landscape and should be given as much thought as the plants you chose—perhaps even more, since they are often permanent.

Hardscapes can be made up of many different materials, the most common being brick, wood, stone, and gravel. Keep your overall design in mind when choosing materials. Although no hardscape will look completely natural, certain materials will blend into a native-plant landscape better than others.

Paths and walkways allow access to inner areas of a garden while reducing the chance for delicate plants to be trampled. Some paths are meant to whisper the way to go; others are meant to shout it. Making a path obvious

Hardscapes are just as important in native landscapes as they are in traditional ones. Choose materials carefully. In most cases you want them to blend into the landscape rather than stand out from it, so stick with earth tones and natural materials when practical.

doesn't mean it has to be boring, however. Bend it around a corner so it disappears for a while, and place plants and stones to break up the site line of the path edges. Avoid edging the path with rigid rows of plants, stones, or logs, and vary the materials that form the path. Meandering paths look more natural than straight ones. Plant your smaller, more delicate flowers along paths where they can be seen and enjoyed.

Paths should blend with their surroundings, but they should always be sturdy and wide enough for safe use. Don't forget to include benches and chairs along your paths and within your gardens so you and your guests have places to sit and enjoy your landscape.

Good walkway materials include brick, gravel, wood chips, and even bare soil. If a path runs under tall pines, use pine needles. If you have a stream or bog garden, you may need to install stepping stones, wooden bridges, or boardwalks.

Steps can be made of stone, natural-looking concrete, or logs. Think of them as terraced elevations formed naturally in the landscape, but be sure that they are safe and easy to negotiate. Wood is slippery when it is wet. Consider installing a handrail alongside your steps. Use small plants and groundcovers to tie the steps into the path and the rest of the landscape.

One of the misconceptions about native-plant landscapes is that they are boring. Unfortunately, sometimes people take native plant design a little too seriously and think they can't use funky garden accents. Not true! There's no reason you can't incorporate any of the garden accents you'd use in a regular landscape. Native-plant landscapes can include sculpture, bird feeders and houses, sundials, gazing balls, fountains, and birdbaths, just like any other garden. It's your garden, and you should include things that make you happy. As with all gardens, keep in mind that these items are meant to be accents and they should be used with discretion and carefully placed rather than just plopped down in the garden.

If you are truly trying to re-create a natural habitat, you'll want to limit your accent pieces to well-placed natural materials such as rocks and logs. Take cues from nature and try to place them as if they had been left there. Moss-covered logs should look like they were once part of a tall forest tree that fell to the ground years ago. Rocks should be buried one-half to two-thirds underground rather than set on the surface, as if a glacier placed them eons ago. Select sculptural pieces that can be nestled into the garden and surrounded by plants, as if they were growing up out of the ground rather than sitting on a concrete pedestal.

Using Native Mosses

One of the best ways to soften rocks or paving stones and make them look as though they've been in place for centuries is with patches of moss. Moss will only grow where soil is shaded and moist with high humus content and a low pH (acidic). Even if you don't have moss on your property, once you set up these conditions, moss will appear, since spores travel for miles.

If you are transplanting moss from another place in your garden, take some soil with it and keep it moist until you are ready to put it in. Before planting moss, work up the soil, freeing it of roots and weeds. Water until it is soaking wet, and allow the surface water to sink in. Then press the patches of moss into the muddy soil, making good contact. Wet the newly planted moss with very dilute fish emulsion. For good measure, pour a cup of milk into each gallon of water you're wetting the moss with to help acidify the soil. Once established in a spot with the right conditions, your moss will need no further care.

Moss will grow naturally where the soil is acidic and moist. Try to look at moss as an asset, and encourage it to grow on fallen rocks and logs to soften them.

Don't be afraid to use nonplant items as accents in native landscapes. These stacked rusty pots are the perfect pedestal for the whimsical fish in this woodland garden.

Nature provides accents in the form of unusual textures and colors. This brightly colored fungus really stands out against the backdrop of fallen leaves in this deciduous woodland.

Using Nonnative Plants

As knowledgeable as you may be about selecting and using native plants, there are still some effects that will be difficult to achieve. The lack of native bulbous plants means you can't get the early, vibrant color from spring bulbs. It will be hard to find a native plant that can give you the season-long show of tropical annuals that so many people rely on for patio containers and hanging baskets. And, there are not a lot of showy vines available for covering trellises and arbors or adding that important dimension of height. If these things are important parts of your landscape design, that's fine. There are many nonnative plants that will readily adapt to the conditions found in a native-plant landscape and complement the native plants.

Most spring bulbs grow well in the same conditions our native prairie plants enjoy, and some even adapt well to the conditions found in a deciduous forest garden. Most minor spring bulbs blend beautifully into spring woodland gardens, bringing colors that aren't commonly found in native plants. These smaller bulbs—such as scillas, snowdrops, dwarf iris, miniature daffodils, glory of the snow, and some species of crocuses—blend into the spring garden well. The bolder colors and exotic look of most tulips, daffodils, crocuses, and hyacinths are too strong for woodland gardens. They can, however, be used in the mixed border with prairie plants.

Bulb foliage should be allowed to die down naturally after flowers fade. Grow them with plants that will mature and fill in while the foliage is still there. Good natives for camouflaging bulb foliage include prairie smoke, columbine, wild ginger, and ferns.

There's no denying the importance of containers in garden design. They bring color and interest to entryways, decks, and patios, and in some situations, are the only way to grow plants successfully. A good way to make containers of nonnative plants blend into your native landscape better is to use containers made of natural materials. Make a path-side planter in a hollowed-out log or stump, or plant a batch of sun-loving annuals in a stone trough.

As a rule, few native plants adapt well to container growing. Many of them have taproots or a limited bloom time. If you do want to use native plants in containers, group several large pots of mixtures of grasses and perennials that will provide interest throughout the season, rather than individual pots containing just one plant. Native plants that adapt well to container culture include little bluestem, rudbeckia, blanket flower, bird's-foot violet, nodding wild onion, and blue giant hyssop.

If your landscape plans includes a vine-covered arbor or trellis and you can't find a vine among the list of natives, several nonnatives are well behaved and blend into native landscapes. Several types of clematis have an understated beauty. Consider *Clematis terniflora* (sweet autumn clematis), *C. texensis* 'Duchess of Albany', or *C.* 'Huldine'. Annual vines such as morning glory, black-eyed Susan vine, and nasturtium will look good with any of the sunny prairie perennials.

For season-long color, it's impossible to beat tropical annuals such as petunias, impatiens, and begonias. Don't be afraid to use traditional annuals in a native-plant landscape. Shade-tolerant types are a good way to bring color into the woodland garden in summer. However, instead of planting one or two six-packs in neat rows, consider a flat or two planted in a large drift—the way nature would do it. Instead of using many different annuals, select one or two types and use them throughout your garden.

In the natural world, you'll find that rich, strong colors such as reds, oranges, and yellows have generally evolved in sunny sites. Cool blues, purples, pinks, and white tend to inhabit lightly shaded areas and woodland gardens. Keep these color schemes in mind when choosing nonnative plants to complement your native plantings and the result will be a much more natural look.

Turf Grasses and Groundcovers

A lawn is a labor-intensive, expensive way to cover the ground. It requires untold hours of mowing, raking, weeding, fertilizing, and watering. This expense and time is justified when you need a resilient ground cover for pets or play areas, but it is an extravagance when the lawn is just a decoration. Maintain only as much lawn as you really use, and replace the rest with a more environmentally friendly and interesting alternative.

There are several alternatives for covering large areas usually given to the traditional lawn. If the area is primarily used for entertaining or relaxing, consider installing a patio or deck and surrounding it with low-growing native plants. If you prefer keeping the area planted, consider a groundcover.

Groundcovers are plants that, by virtue of creeping runners, stolons, or rhizomes, form a low, self-spreading carpet of greenery that prevents erosion and do not need to be mowed. However, ground covers do not take foot traffic. Several native-plant nurseries have started offering lower-maintenance alternatives to traditional Kentucky bluegrass lawns for situations where people don't want

The delicate purple of scillas is a perfect complement to early-blooming woodland plants. This minor bulb can tolerate the shade found under deciduous trees, flowering and storing the necessary energy before the trees fully leaf out.

Few native vines adapt well to use in formal landscape settings. Fortunately, there are several well-behaved exotic species, including *Clematis* 'Huldine', that combine nicely with native plants.

"Low-mow" mixes are becoming popular as alternatives to traditional lawn grasses. They are good choices in well-drained areas with minimal foot traffic and partial to full sun.

Responsible Lawn Care

In the areas where you do want to have lawn, practice responsible lawn care. Here are some tips for minimizing the negative effects of large areas of lawn grasses.

❖ Start with a soil test to determine which, if any, nutrients are lacking. There's no need to apply fertilizer if your soil has ample nutrients. Over-fertilized soil encourages excess growth and runoff.

❖ Add ½ inch of compost twice a year in spring and fall to provide necessary nutrients.

❖ Mow at 3½ inches to 4 inches. Longer grass shades its own roots, reducing water loss. Taller grass is also better able to outcompete many annual weeds including crabgrass.

❖ Don't remove more than one-third of the blade at any mowing.

❖ Let clippings fall to recycle nutrients. They do not lead to thatch problems.

❖ Water wisely. Most lawns are over-watered. Allow your cool-season lawn grasses to follow their natural cycle and go dormant in mid to late summer. They will green up again once the cooler, wetter days of late summer roll around.

to fuss with a highly manicured lawn but still want some kind of low-growing, green play area. These are usually mixtures of slow-growing fescues that will form a soft 4- to 6-inch-tall carpet of grass. They can be mown once a month to a height of 3 to 4 inches. These mixes grow best in sun or partial shade in loamy or sandy soils and are not recommended for wet soils, deep shade, or clay soils that exhibit standing water after a rain.

Native Groundcovers

Many native plants have a tendency to expand and fill in areas, eventually covering the ground. Typically, a groundcover plant is low-growing and aggressive to the point of keeping others out and does not require a lot of care once established. Here are some good choices:

Anemone canadensis (Canada anemone)
Antennaria species (pussytoes)
Arctostaphylos uva-ursi (bearberry)
Asarum canadense (wild ginger)
Comptonia peregrina (sweet fern)
Euphorbia corollata (leafy spurge)
Fragaria virginiana (wild strawberry)
Heuchera species (alumroots)
Juniperus species (junipers)
Maianthemum canadense (Canada mayflower)
Onoclea sensibilis (sensitive fern)
Opuntia species (prickly pears)
Osmunda species (ferns)
Viola species (violets)
Waldsteinia fragarioides (barren strawberry)

Dealing with City Laws and Neighbors

Homeowners who decide to go native are often confronted with an array of laws, regulations, requirements, and sometimes outright hostility. Most of this is simply due to ignorance, which is difficult to overcome, at least at first.

There are several things you can do to prepare yourself for neighborhood opposition. To begin, learn your local laws and ordinances. Chances are you will be able to stay well within them by putting some thought into your landscape before you plant.

Once your landscape is planted, be attentive and keep up with the necessary maintenance. Let your neighbors know by your presence and activities that your yard is being cared for and not neglected. Be purposeful. Talk with neighbors about your plans before you start planting, if possible. Educate. Take the time to tell your neighbors what the plants are and what butterflies and birds they attract. Share plants and seeds with them if they are interested. In the end, you may make a new convert.

Adiantum pedatum (maidenhair fern) and *Mertensia virginica* (Virginia bluebells) are good choices for more-formal landscapes in shade.

Prunus nigra 'Princess Kay' has a "cultivated look" and adapts well to front-yard use.

Front-Yard Native Plants

Most native plants will stay neat and tidy if properly pruned and groomed. However, some are easier to keep looking neat than others. Here are some easy-to-establish, attractive native plants that adapt well to more-formal landscape situations without a lot of attention. Keep in mind also that cultivars, when available, often look less wild than the species.

Herbaceous Plants for Sunny Sites

Agastache foeniculum (blue giant hyssop)
Allium cernuum (nodding wild onion)
Asclepias tuberosa (butterfly weed)
Aster laevis (smooth aster)
Aster novae-angliae cultivars (New England aster)
Boltonia asteroides (boltonia)
Echinacea species (purple coneflowers)
Euphorbia corollata (flowering spurge)
Gaillardia species (blanket flowers)
Geum triflorum (prairie smoke)
Helenium autumnale (autumn sneezeweed)
Heliopsis helianthoides (oxeye)
Liatris species (blazing stars)
Lupinus perennis (wild lupine)
Monarda species (wild bergamot, bee balms)
Panicum virgatum (switch grass)
Phlox pilosa (prairie phlox)
Physostegia virginiana (obedient plant)
Pulsatilla patens (pasque flower)
Ratibida pinnata (gray-headed coneflower)
Rudbeckia hirta (black-eyed Susan)
Schizachyrium scoparium (little bluestem)
Sporobolus heterolepis (prairie dropseed)
Verbena stricta (hoary vervain)
Zizia aptera (heart-leaved alexanders)

Herbaceous Plants for Shady Sites

Actaea species (baneberries)
Adiantum pedatum (maidenhair fern)
Aquilegia canadensis (Canada columbine)
Aralia racemosa (American spikenard)
Arisaema triphyllum (Jack-in-the-pulpit)
Asarum canadense (wild ginger)
Athyrium filix-femina (lady fern)
Claytonia virginica (Virginia spring beauty)
Dicentra species (Dutchman's breeches, bleeding hearts)
Erythronium species (trout lilies)
Geranium maculatum (wild geranium)
Hepatica species (hepaticas)
Isopyrum biternatum (false rue anemone)
Lobelia cardinalis (cardinal flower)
Maianthemum canadense (Canada mayflower)
Matteuccia struthiopteris (ostrich fern)
Mertensia virginica (Virginia bluebells)
Phlox divaricata (blue phlox)
Polemonium reptans (spreading Jacob's ladder)
Polygonatum biflorum (giant Solomon's seal)
Polystichum acrostichoides (Christmas fern)
Sanguinaria canadensis (bloodroot)
Solidago flexicaulis (zigzag goldenrod)
Trillium species (trilliums)

Woody Plants

Abies balsamea (balsam fir)
Acer rubrum (red maple)
Acer saccharum (sugar maple)
Amelanchier × grandiflora (apple serviceberry)
Betula nigra (river birch)
Carpinus caroliniana (blue beech)
Clematis virginiana (virgin's bower)
Cornus species (dogwoods)
Dirca palustris (leatherwood)
Gymnocladus dioica (Kentucky coffee tree)
Ilex verticillata (winterberry)
Juniperus species (junipers)
Ostrya virginiana (ironwood)
Physocarpus opulifolius cultivars (ninebark)
Potentilla fruticosa cultivars (shrubby cinquefoil)
Prunus nigra 'Princess Kay' (Canadian plum)
Quercus species (oaks)
Tilia americana (basswood)
Tsuga canadensis (eastern hemlock)
Viburnum species (viburnums)

Chapter 5

Gallery of
Gardens

A Collection of Minnesota Native Plant Gardens

Backyard Oasis

A Small City Garden Provides Respite for Birds and Butterflies

Robert and Marlene Olsen must have done something right while raising their son Erik. When bird and butterfly enthusiast Marlene said she'd like to see a wider variety of species in her backyard, Erik offered to design a garden for her. Erik Olsen, a landscape designer-MLA with Tennant Landscaping, Inc., used his knowledge and enthusiasm for native plants to create a backyard habitat full of birds and butterflies for his mother, and also captured the heart of his somewhat reluctant father.

Before 1997, the Olsen backyard was like every other backyard in their Saint Paul neighborhood, a monoculture of Kentucky bluegrass. Erik used a sod-stripping machine to remove the bluegrass, and was pleased to find the soil was an organic, sandy loam soil perfect for mesic prairie plants. Recalling visits to remnant prairies and savannas, Erik incorporated pussy toes, wild petunia, great lobelia, purple prairie clover, blue giant hyssop, little bluestem, prairie smoke, blazing stars, butterfly weed, wild bergamot, purple coneflower, New England aster, sky-blue aster, columbine, and prairie dropseed. To keep costs down, he planted with small plugs, 4-inch potted plants, and 2- and 5-gallon container shrubs. He kept the bare soil areas covered with a 1-inch layer of cocoabean mulch the first three years to reduce weeding and watering needs. Now that the plants are established, the only mulch used is shredded pine bark on some of the shrubs, which include American hazelnut, snowberry, 'Regent' serviceberry, New Jersey tea, leadplant, and sand cherry.

Now that the garden is established, maintenance needs are minimal. Erik spends a day each spring cut-

Butterflies flock to the *Asclepias tuberosa* (butterfly weed) and *A. incarnata* (swamp milkweed), two of their favorite nectar sources.

ting back the herbaceous plants. After that, his parents can get by with what he calls the "15-minute-a-day plan" for weeding. They spend 15 minutes or so in the garden each day in spring getting weeds under control, and by the end of June, it is down to even less time. Their biggest weed challenge has been with creeping bellflower (*Campanula rapunculoides*), a European plant that has invaded many Saint Paul gardens. The only watering necessary is for newly planted seedlings, which Erik marks with a stake so his parents can find them among the larger mature plants. Rabbits have caused some damage on eastern sand cherry. The Olsens protect young plants with chicken wire.

Erik is just as pleased with the results as his parents are. He said the only thing he wouldn't have done was include goldenrods in the original planting plan. Not because he doesn't appreciate this group of natives; they just seed a little too prolifically for a small garden. Erik, who prefers the title "horticultural ecologist," is quick to credit the natural world for his designs. He says a visit to a native habitat is the best way for him to learn how to arrange natives in combinations that work in typical landscape situations. Judging by the lovely garden he's created for his parents and for the birds and butterflies it attracts, it's obvious his parents and nature have both taught him well.

Native grasses and asters provide interest in fall.

In mid-May the garden is full of color from *Aquilegia canadensis* (Canada columbine) and *Geum triflorum* (prairie smoke).

Erik marks newly planted seedlings so his parents can find them among the larger mature plants and water them.

61

The Best of Both Worlds
A Woodland Retreat Blending Native and Nonnative Plants

It's difficult for Mary Stanley to pinpoint when her love of wildflowers began. She remembers growing up in the country outside Milwaukee, Wisconsin, and walking through the woods and fields looking for wildflowers. While living in Milwaukee in the late 1960s, she become involved with the Riveredge Nature Center, where she volunteered with native-plant guru Lori Otto and became a teacher-naturalist at the center, spurring her interest in native plants. She even took courses in garden design and began designing gardens for other people. When she moved to Minnesota in the 1980s, she decided she wanted to devote her time to her own garden and her love of native plants.

It was only natural that when it came time to select a lot for their new house, Mary and her husband, Dick, were drawn to a heavily wooded site in Washington County. Even before their house was built in 1989, Mary was planting wildflowers on the lot. She was blessed with a builder and construction crew who were sensitive to her desire to preserve the natural features of the lot. Mary went so far as to wrap construction tape around the large trees she wanted to protect. Her persistence paid off. Not one tree was damaged during construction, and today the house is nicely framed by the large oaks and maples.

Mary tackled her landscape in stages, putting the highest priority on the knoll area out front. She describes the soil as a well-drained glacial till. Her front-yard garden is home to many spring ephemerals, including trilliums, Dutchman's breeches, bloodroot, and spring beauties. As showy as they are in spring, a front yard of ephemerals can become pretty bleak by midsummer, unless you know what to plant with them. And, this is where Mary really shines. She has a knack for combining native flowers with natural-looking nonnatives that adapt well to the same conditions. She stays away from hostas, which she feels are too coarse, but likes epimediums and primroses. She has found the Japanese counterparts to our native species to be effective. She likes *Anemonopsis macrophylla*, a summer-blooming woodland plant with waxy, lavender flowers; *Kirengeshoma palmata*, a zone 5 plant growing 3 feet tall with yellow nodding flowers in late summer; and *Glaucidium palmatum*, a zone 5 woodland treasure with poppylike flowers for three to four weeks in spring.

Once Mary had the front garden well underway, she moved around to the back of the house. Because the architect successfully utilized the natural topography to site the house on the hillside, Mary was left with nine layers of exposed cinder block on one corner. She had some of the heavy clay dredged out of the nearby pond and created a hillside garden against the house. Large boulders were brought in to create a tiered rock garden and retain the slope. As with the building of the house, Mary was hands on, making sure each boulder was properly placed and buried deep enough to look natural. Even though this garden is on the north side of the house, it is the sunniest part of the yard. Here, Mary plants her "treasures," including some things she has saved from her mother's garden.

Because many of the plants Mary loves can be difficult to find, she has become adept at starting plants from seeds. It's not always the fastest route to get a plant in the garden, but in some cases it's the only way. Mary has a batch of trilliums she started from seeds. After four years in the cold frame, the plants finally sprouted a single leaf. She's hoping to see blooms in another three years or so.

Large oaks and maples provide the canopy for the front yard, where spring ephemerals and shade plants carpet the ground.

The key to Mary's success with native plants has been her method of soil improvement. It's intense, but it has paid off. Every fall she rakes off all the leaves, shreds them, and them puts them back on the gardens to return the nutrients to the soil. It's not always easy racing against Mother Nature, and some years she ends up waiting until spring to get all the persistent oak leaves shredded, but she always manages to get the task done.

Once the beds are mulched, maintenance is minimal. Mary still pulls an occasional buckthorn seedling, and the combination of shallow maple roots and well-drained soil means she has to water the front area during dry periods. One thing she never has to do is mow the lawn. Mary prides herself on the fact that her landscape has never had a blade of turf grass. Even along the road, Mary has planted sun-loving native prairie plants such as wild lupine, prairie smoke, golden alexanders, and assorted grasses. Her landscape is a wonderful example of how to complement the natural beauty of native plants with well-behaved exotics.

The Stanleys' home fits naturally into their site. The back of the house features the hillside garden and a pond.

In late May, color in the front woodland comes from *Trillium grandiflorum* (large-flowered trillium) and *Phlox divaricata* (woodland phlox) against a backdrop of *Podophyllum peltatum* (May apple).

To continue the show into summer, Mary uses natural-looking nonnatives such as lungworts (*Pulmonaria* species), which adapt well to the same conditions as her woodland natives.

Persistence Pays Off
Determination Brings Well-Deserved Rewards

Barbara Pederson would never say her native-plant landscape has been easy, but she will always tell you that it has been worth it. When she and her husband, Don, purchased their North Oaks golf course property in the mid-1970s, it was heavily wooded, the large elms and birches completely screening the view of the course. However, by the early 1990s, most of the elms had succumbed to Dutch elm disease. Although Barbara welcomed the newly opened vista provided by the golf course, she wanted something more interesting than a backyard of turf and started looking for other options for the 1-acre site. One nurseryman suggested an apple orchard, but that seemed like too much work and offered little appeal. When Barbara asked him about wildflowers, he said he had little experience with them. Barbara learned that the North Oaks Garden Club was experimenting with native plants, and one member suggested she contact Native Heritage Landscapes for help.

Site preparation began in fall 1993 and included several applications of Roundup, letting the vegetation grow up, and spraying again. Planting did not take place until summer 1995, and included a seed mixture of sixty species of forbs and eight species of grasses, as well as 500 small plugs of flowering plants. A bluegrass path through the garden was also seeded at this time.

For the first five years, Barbara tended the area herself, the only professional help coming in the form of a spring burn each year. However, weedy plants were becoming a problem, and by 2000, thoughts of abandoning the project were crossing her mind. She decided to seek the help and advice of several local native-plant experts. She first contacted Chase Cornelius, a volunteer who worked with the prairie at the Minnesota Landscape Arboretum. He suggested choosing target areas to concentrate her weed-control efforts and then replanting with larger plants (in 4-inch pots) to better compete with the weeds. Ann Mueller and her crew at Go Native weeded one large target area by hand and then laid thick layers of newspapers covered with 2 to 3 inches of shredded bark to smother the weeds. Barbara says this area is still relatively free of weeds. She also got help from Ron Bowen and Beth Fritz of Prairie Restorations. Beth designed the attractive center garden within the prairie area, an area about 20x20 feet that features gravel mulch and a rock wall that supports low-growing forbs and grasses. Prairie Restorations has been assisting Barbara with maintenance since then, including burning or mowing annually.

Barbara also has a woodland garden that she started in the 1990s. Located along one side of the prairie, it is shaded by mature pines, birches, and amur maples. Plants that do well here include Jack-in-the pulpit, blue phlox, wild geranium, violets, Virginia bluebells, and maidenhair fern. The small plot size of approximately 50x25 feet has proved to be too much for ostrich ferns and snakeroot, which require regular weeding out. Maintenance is minimal in this garden, mainly focused on removal of buckthorn seedlings.

Despite the many challenges Barbara has faced, not only with the weeds but also with deer, woodchucks, and rabbits, she says it has been a joy watching the gardens develop. She loves listening to the songs of the birds and watching the monarch butterflies flittering around while she tends the garden. She enjoys watching the parade of blooms throughout the season, starting with golden alexanders, blue phlox, and wild lupine, and continuing with purple coneflower, butterfly weed, great blue lobelia, cardinal flower, New England aster, blazing stars, and ironweed. Her prairie garden has turned out to be the perfect complement to the uninterrupted pastoral backdrop of the golf course.

Purple-flowered *Liatris pycnostachya* (great blazing star) and *Veronica stricta* (hoary vervain) provide accent to the mass planting of *Rudbeckia hirta* (black-eyed Susan) that surrounds the center gravel garden.

When viewed from the house, the prairie garden provides the perfect transition to the golf course.

The attractive center garden features gravel mulch and a rock wall that supports low-growing forbs and grasses.

In late August the native grasses start to take on their autumn colors, giving the garden a softer, earthier look.

65

Haven for Wildlife

A Pond Surrounded by Native Plants in a Suburban Sanctuary

When Peggy and Wayne Willenberg were looking for property for their new house, they didn't just consider their needs. They also took into consideration the needs of any wildlife that would potentially share their home. Peggy's lifelong interest in nature led them to a 5-acre property in Plymouth. Although it was not much more than an open field with a small woodlot, Peggy saw the potential. Since moving in, the Willenbergs have developed their property to the point where it has become certified by the National Wildlife Federation as an official Backyard Wildlife Habitat.

The front of the property is a 1½-acre pasture providing shelter for nesting birds and other wildlife. The backyard includes a restored prairie and a more formal stroll garden featuring native plants attractive to wildlife. However, the real attraction of the property—for both humans and wildlife—is the pond, which is surrounded by a deciduous woodland garden. Meandering pathways and benches provide welcome respite for people, while the pond and plantings provide sustenance for fish and waterfowl, frogs and toads, minks and muskrats, and larger, four-footed visitors.

The pond was created in a low area toward the back of the property. Peggy wanted a natural-looking water feature that would be as attractive to wildlife as it was to people. She got design help from landscape architect Diane Hilscher and installation and maintenance help from Joni Cash of Cash & Roberts. Together they developed a plan that included not only a pond but also a stream, waterfall, and woodland garden. The pond, which has a natural clay bottom and is no more than 5 feet deep, was dug out with a Bobcat in 1992. Because they weren't able to start planting right away, it quickly became overgrown with cattails. Although this might have been an attractive habitat for wildlife, it wasn't suitable for people, so they set about the challenging task of removing the cattails, which included burning, spraying, and cutting. It took two years to completely eradicate the cattails and get to a point where they could begin planting.

When it came time to select plants for the pond, Peggy turned to the garden of her favorite impressionist, Monet. She wanted to capture the way he wove things together to create a naturalistic look. She incorporated the pinks and purples Monet used, substituting plants suitable for a Minnesota garden. Most of the plants around and in the pond are native, but she has also had to use some nonnatives to get the effect she was looking for, including hardy rhododendrons, *Astilbe chinensis* 'Visions', and weeping willow. The pond is home to water lilies, pickerelweed, and arrowhead. Alongside the pond, moisture-loving plants such as Culver's root, marsh marigold, blue flag, swamp milkweed, obedient plant, and white turtlehead give the desired effect. In the nearby

Wet-soil plants planted along the edge of the pond create a smooth transition to the adjacent prairie.

woodland garden, baneberries, wood anemone, Jack-in-the-pulpit, wild ginger, sedges, ferns, May apple, blue beech, and a large planting of yellow lady's slippers weave together. Wild onion, New England aster, Joe-pye weed, shooting star, and bottle gentian help create a smooth transition from the pond to the adjacent prairie.

Peggy also wanted to duplicate the tranquil effect created by the fish found in Monet's water garden, so she purchased a few Koi for her own pond. However, it quickly became obvious that Koi were a favorite food of mink. Not only did mink now make frequent summer visits, but they even considered the pond their own "fish refrigerator" in winter, reaching in through the unfrozen section for a quick snack. Although this fulfilled Peggy's goal of attracting wildlife, it was a little too expensive to keep restocking Koi just to keep the minks fat and happy. She now buys small feeder goldfish every year, which become huge by the end of the season, and happily shares some with the mink, egrets, and the other wildlife that are attracted to the fish.

The Willenberg property has truly become a haven—both for people and for wildlife. Peggy says she and her family get immeasurable enjoyment not only from the plants, but also from the endless stream of wildlife that keeps them entertained year round, from the spring concerts of the male frogs to the winter visitors that come to the pond for a drink.

A stepping-stone path winds its way through the woodland garden behind the pond.

The large weeping willow helps give the pond the Monet look Peggy was looking for.

A pond-side bench provides a place to sit while waiting for wildlife to visit the Willenbergs' garden.

This waterside planting includes *Sagittaria latifolia* (broad-leaved arrowhead), *Lobelia siphilitica* (great lobelia), and *Actaea rubra* (red baneberry).

A restored prairie is part of the backyard.

A Prairie Paradise

A Suburban Home Enveloped by the
Magic of a Restored Prairie

Most people move to the country to surround themselves with the shade and seclusion of large shade trees, but Pat and Bob Angleson bought their Grant Township property because they wanted to embrace the openness. Large elms shaded their city lot, and they were looking forward to enjoying sunlight and open space. They chose a site with rolling hills and sparse trees that would have once been home to a thriving oak savanna habitat.

Once the construction of their prairie-style house was underway in summer 1993, the Anglesons turned their attention to landscaping the large site. Pat admits that she enjoys doing yard work, but 3 acres of mowing was too much for even her, and she was concerned about the time and cost of maintaining such a large lawn area. Pat remembered how much she enjoyed walking through the virgin patches of prairie on her parents' land in Roberts, Wisconsin, and she decided that was what she wanted to do with her 3 acres.

Pat's first challenge was finding knowledgeable people to help her with her prairie project. You couldn't just pick up the phone book and find a list of native-plant specialists in the early 1990s. She started making inquiries and found Sonja Moseman, owner of Native Heritage Landscapes, who helped her get started.

Excellent site preparation has been the key to success in the Angleson prairie. By taking the time to try to exhaust not only the weeds but also the existing seed bank, they reduced their weed problems later on. The first step was to mow the entire field and burn off the residue.

The restored prairie is full of color in early August, with *Monarda fistulosa* (wild bergamot), *Ratibida pinnata* (gray-headed coneflower), *Agastache foeniculum* (blue giant hyssop), *Heliopsis helianthoides* (oxeye), *Echinacea purpurea* (purple coneflower), and *Eryngium yuccifolium* (rattlesnake master).

When new vegetation sprouted, they sprayed Roundup. After a few weeks, they plowed under the residue and allowed plants to regrow so they could spray a second application of Roundup. They plowed under this round of residue and waited for another batch of seeds to sprout. A final Roundup application killed off the next round of vegetation, but they did not plow a final time, fearing that would turn up too many weed seeds. They seeded a general prairie seed mix in spring 1995.

The first year, Pat admits things looked "pretty ragged." She remembers the neighbors laughing at her crop of thistles, mulleins, and dandelions. The next summer they mowed the entire 3 acres, and that's when the grasses really started to appear. Pat says she was redeemed in her neighbors' eyes by the large showy areas of rudbeckia that second year. Things have become more balanced every year since then.

Pat loves the prairie because of its diversity and seasonality. It changes subtly every day, and every year different plants are highlighted, unlike a static typical home landscape where once things are planted, they pretty much stay the same. One year false indigo will be the star; the next year coneflowers or purple prairie clover will take center stage.

Now that the prairie has been around awhile, it has reached a sort of equilibrium and requires little maintenance. It has been drought tolerant and disease resistant, and Pat says she never sees serious insect damage. She occasionally weeds out aspen seedlings, is always on the lookout for knapweed, and occasionally does some spot control of sweet clover. Dandelions appear on the paths, where they are easily mown down, but they haven't been a problem in the prairie itself. Pat gets maintenance help from Prairie Restorations; she says the staff does an excellent job of helping her monitor and tackle problems before they get out of control. They are keeping an eye on a patch of reed canary grass that is encroaching on one end. Pat is mowing it regularly in hopes of keeping the plants weak and less aggressive. If it becomes a problem, Prairie Restorations will spray the area with Roundup and do some slit seeding.

Pat is hard pressed to think of any drawbacks to her prairie. The only thing she could come up with was that it takes so little care now that it's established. Being a gardener, she admits feeling a bit frustrated at times because she can't really pull stuff out and replant like a normal landscape. She satisfies her need to tend plants by keeping a more-cultivated garden area around her house. The plantings here aren't all prairie plants, but they were chosen (with the help of landscape designer Diane Hilscher) to echo the native plants of her prairie.

Wild lupines turn the prairie into a sea of blue in late May.

The Anglesons try to burn their prairie every three or four years to help keep woody plants from encroaching as well as to enrich the soil.

Above: The 3-acre prairie in their front yard sets off the Anglesons' prairie-style house nicely.

Left: A walk through the prairie reveals small vignettes within the large expanse. Here *Rudbeckia hirta* (black-eyed Susan), *Dalea purpurea* (purple prairie clover), and *Verbena stricta* (hoary vervain) bloom together.

71

Successful Sustainability
A Little Work Up Front Resulted in Success for This City Landscape

After doing extensive remodeling to his Victorian-era home, Philip Friedlund turned his attention to the surrounding landscape. Although the previous owner had been a gardener, the perennial plantings had become overgrown and no longer matched the site and what Philip wanted from the landscape.

Philip's ideas for his landscape renovation were inspired by several visits to Sweden in the late 1980s. He was impressed with the Swedes' knowledge of wildflowers and how they used them freely in their landscapes. In 1990, he contacted the University of Minnesota's Landscape Architecture Department to find someone who could help put his ideas into place, and he was connected with Fred Rozumalski, a student and avid native-plant enthusiast who was just launching his career as a landscape architect. Fred convinced Philip to create not only a front-yard garden, but to redo the entire lot featuring native plants. Their goals were to create a sustainable landscape that would not require intensive maintenance or supplemental watering, fertilizing, or pesticides, a landscape that would accommodate wildlife, be species diverse yet visually unified, and provide year-round interest.

The first step was to reduce the size of the lawn, keeping only what was needed for daily use and to accommodate circulation. However, because of the home's prominent position in the Ramsey Hill neighborhood of Saint Paul, they needed to be cognizant of how a native-plant landscape might appear to uneducated neighbors and passersby. Philip talked with neighbors and got their support, and Fred incorporated traditional elements into the design, such as crisp lines between native plantings and turf, lots of flowering native species, and the creation of pattern within the native plantings to give a sense of intent and order.

The majority of the plants in the landscape are Minnesota natives, with some noninvasive, drought-tolerant, exotic perennials filling in as needed. The sunny front

Philip Friedlund has successfully surrounded his Victorian-style St. Paul home with a native landscape.

To give a sense of intent and order, the landscape includes traditional elements such as crisp lines between native plantings and turf, pathways, and a sundial.

yard is filled with native prairie plants. Spring color comes from prairie smoke, blue-eyed grass, and heart-leaved alexanders. Butterfly weed, liatris, purple prairie clover, monarda, and prairie phlox take center stage in summer, followed by grasses, ironweed, coneflower, asters, goldenrods, and Joe-pye weed.

Mature trees shade the backyard, so Fred looked to the shaded microclimate of the oak savanna and included Canada columbine, wild geranium, bloodroot, false Solomon's seal, and an assortment of ferns. Since these plants tend to be spring bloomers, Fred supplemented this area with a few summer-blooming exotic plants, including *Astilbe chinensis* 'Pumila', *Bergenia cordifolia*, *Dicentra spectabilis* (bleeding heart), and *Pulmonaria saccharata* 'Sissinghurst White'.

The shadier backyard is a perfect spot for *Aquilegia canadense* (Canada columbine) and *Geranium maculatum* (wild geranium).

Another goal of the design was to hold as much rainwater (and therefore nutrients and particulate matter) on the site for as long as possible to encourage infiltration into the soil. Fred accomplished this in two ways. First, shallow water-detention depressions (rainwater gardens) were created and planted with a variety of prairie plants that thrive in moist soil. Second, the lawn area was slightly dished to encourage water infiltration and reduce the need to water.

Maintenance of the Friedlund landscape is greatly reduced from that of a more traditional landscape. Philip and his wife, Lisa Isenberg, start cutting some plants back in fall, but for most plants they wait until early spring, to take advantage of the wonderful display of color and texture in winter. Plant refuse goes into the compost pile to be recycled for mulch and as a soil amendment. Some hand-weeding is needed, and the front yard does require watering during dry periods to maintain the health of the lawn and some perennials.

The Friedlund landscape has been a success on all accounts. Not only does the entire family enjoy the garden, but it's also a source of pleasure for anyone who passes by. Philip says the garden has become a destination of sorts—a place people purposefully include in their daily walks or commutes—and he couldn't be more pleased.

Late-summer interest comes from *Eupatorium maculatum* 'Gateway', *Panicum virgatum* (switch grass), and *Ratibida pinnata* (gray-headed coneflower).

Flowers and Groundcovers

Native flowers and groundcovers are the funnest aspect of landscape design. Their showy blooms, various sizes and forms, and interesting leaves add color, interest, and texture. Many are attractive to butterflies, birds, and bees, and some are good for cutting. Others are more functional, offering erosion control on a slope or an attractive way to deal with a low area in the lawn or dry shade under large trees.

Many native flowers are well suited to garden and landscape use. Some you may already know and grow. Wild bergamot, butterfly weed, Joe-pye weeds, asters, and black-eyed Susan are all commonly grown in perennial beds and mixed borders, and wild ginger, bloodroot, and blue phlox are common in shade gardens. Many other native flowers adapt well to cultivation, especially if you can provide them with suitable soil and site conditions.

Groundcovers are plants that spread readily to cover the ground, usually in tough sites. Typically these are low-growing plants, but some are a foot or more tall. Most grow in shade, but some need sun. Groundcovers take a little effort to get established, but once they are set, they can be an effective way to reduce mowing and other maintenance.

When choosing flowers and groundcovers, think in terms of ecosystems instead of individual plants. If you have a sunny area with well-drained sandy soil where you want to put a flower garden, look to the dry-prairie ecosystem. Choose plants that are native to this habitat with a wide variety of bloom times and you'll end up with a showy, low-maintenance border with interest from spring until fall. If you have an area under the canopy of large deciduous trees, look to the shade-loving plants of the deciduous forest. By planting flowers and groundcovers that have evolved together naturally, you'll end up with a group of plants that not only grow well together, they'll look good together too.

Planting

There are several ways to plant flowers and groundcovers. Seeds of native plants are available from many specialty nurseries, both as individual species and as mixes. Some plants, especially woodland types, require a cold-moist period or some type of mechanical scarification to break dormancy. If you buy your seeds from a knowledgeable source, you'll get detailed germination information. Many native plants are available bare-root from mail-order nurseries and container-grown from nurseries and garden centers. New plants can also be obtained by divisions of older established plants that you have growing in another area of your landscape, or from a friend. Never dig plants from the wild.

If you will be planting a large prairie garden, a seed mix is the most practical way to go. Seeds take longer to become established plants and reach flowering age, but the result is a more natural look and it is more economical. Order seed mixes from a nursery specializing in wildflowers in your area. If your goal is a true prairie restoration, try to find a nursery within about 200 miles of you to ensure an authentic genotype. Be wary of wildflower-in-a-can mixes. Many contain nonnative plants, including annuals that won't be around for more than a year or two. Prairies can be seeded in spring or fall, but to be successful, seeds must be sown on a well-prepared seedbed, not just scattered in existing vegetation. You can augment your seeding with plants to get flowers faster.

Bare-root plants must be planted in early spring. Dig a hole slightly deeper and wider than the longest roots of your plant. Make a cone of soil in the center of the hole to set the plant on. The crown (where the stems join roots) should be level with the soil surface. Spread roots out evenly and refill the hole with soil, watering well. Spring is also the best time to plant most container-grown native flowers, but they can be planted throughout the growing season if you pay close attention to watering. Choose a cloudy day to plant, and give all newly planted flowers a thorough drink of water.

Most native flowers and groundcovers can be divided in spring or early fall. Water thoroughly and keep the soil moist until plants are established. It is a good idea to put down a winter mulch of weed-free straw or leaves after the ground has frozen. This helps ensure that the new plants will remain firmly planted in the soil through winter freeze-thaw cycles.

Space plants based on their mature sizes and their use. Roughly, perennials should be spaced half their height apart. If they grow 2 feet tall, place them about a foot apart. Groundcovers should be spaced a little closer than is recommended so they will fill in faster.

Although some flowers can be used as single specimen plants, most flowers look best in groups of three, five, or seven or more, as they are in nature. Repeating a plant or a grouping is an effective technique in any garden where there is sufficient room.

Many native flowers are ephemeral in nature, meaning they go dormant during summer's heat. Plant them with summer plants that will cover the bare ground when the ephemerals "disappear." Be careful not to dig up dormant clumps of spring ephemerals. If you will be cultivating that area of the garden, you may want to mark them in late spring so you don't disturb their roots in summer or fall.

Many native flowers and groundcovers readily adapt to landscape use. Two of the most popular are *Rudbeckia hirta* 'Goldsturm' and *Eupatorium purpureum* 'Gateway', growing here with *Echinacea purpurea* (purple coneflower) and *Heuchera* 'Palace Purple' in a perennial border.

Care

Sunny prairie plants grow in almost any average garden soil, but most shade-loving native flowers and groundcovers do best in slightly acidic, well-drained soils with at least 40 percent organic matter. Whenever possible, add organic matter to the soil before planting any native plant. Dig the garden bed as deeply as possible and incorporate 3 to 4 inches of organic matter—well-rotted manure, compost, or peat moss. Some plants require nitrogen-fixing bacteria, which can be incorporated into the soil in the form of a commercially available inoculant. Even though many native plants tolerate tough conditions, most will grow better in a garden setting if the soil has been amended with organic matter, including dry-soil prairie plants.

Flowers and groundcovers will benefit from a 1- to 2-inch layer of organic mulch to retain moisture, keep weeds down, and improve the soil. Good mulches are shredded leaves, pine needles, or compost. Shredded bark can be used in woodland gardens and mixed borders, but it's usually too coarse for perennial beds.

If you've amended the soil with organic matter before planting and you keep plants mulched, most native flowers and groundcovers will not need additional fertilizer. If you do want to fertilize, in early spring or fall apply a layer of well-rotted compost around plants. You can also use an organic fertilizer such as Milorganite, fish emulsion, or blood meal. Apply these in early spring, and water well.

As with all garden plants, water is important for establishment during the first year. Once plants are established in an appropriate site, a layer of mulch should be all most plants need. Prairie plants especially prefer to be on the drier side. If you are incorporating new plants into an established bed, use a stake or some other type of marker to remind you to give them extra water.

Weeds are always a problem for gardeners, even in established landscapes. The best way to keep weeds under control is a good defense. Kill weeds before planting, if possible. Once your garden is planted, keep the bare soil covered with mulch. Try to pull or cut back annual weeds before they go to seed. Avoid using pre-emergent herbicides, as they will prevent all seeds from sprouting, including self-sown seedlings of your native plants that give your garden the natural look you want.

There are other maintenance tasks you may choose to perform on some flowers. Taller plants may need staking or supports to keep them from flopping over. In prairie gardens, support tall flowers by placing them near stiff grasses such as big bluestem and Indian grass. Use wooden stakes or small tomato cages to support individual plants.

Most later-blooming plants can be pinched or cut back by about half in late May, or at the latest, early June. This keeps the plants shorter and more compact. However, if you do it too late, you run the risk of no flowers. Deadheading is the cutting out of spent flowers to encourage more bloom. It is more important with nonnative annuals, but it does help some natives produce more flowers. However, some native flowers, such as blazing stars and coneflowers, have seed heads that are attractive to birds and that also offer winter interest; these seed heads should not be removed until spring.

Possible Problems

Most native flowers and groundcovers will not be bothered by serious insect or disease problems if they are growing in appropriate sites. Keep moisture-loving plants in heavier soils and dry-soil plants out of low areas, and match plants to available sunlight conditions. Powdery mildew is a fungal disease that may show up during wet summers on some plants, but it's rarely serious. Cut back diseased plants in fall and remove the foliage. If the problem continues year after year, choose another plant for that site.

Common insect pests include slugs and leaf miners, both usually resulting only in cosmetic damage. If you want to have butterflies and other beneficial insects in your garden, you'll have to learn to live with a few holes in leaves now and then. If you find a heavily damaged plant, put in a few more plants to compensate for what you'll be sharing with the butterflies. Whatever insect problems you face, don't resort to insecticides. They kill too many beneficial insects, including bees and butterflies.

Actaea rubra
Red baneberry
Zone 2

Actaea rubra

Native Habitat Woodland areas with rich, neutral to acidic soils throughout Minnesota.

Height 12 to 30 inches

Description Showy clusters of dense, fluffy white flowers rise above attractive deeply cut compound leaves starting in late April and continuing well into May. Shiny red berries follow flowers in late summer. Mature plants have a shrublike appearance.

Landscape Use Red baneberry is well suited to woodland or shade gardens, where it will be an attractive addition throughout the growing season. Its shrublike appearance allows it to be used in foundation plantings. Place it where the showy red fruits will be enjoyed in late summer; a backdrop of lacy ferns is nice. The berries are poisonous to people, but are a favorite food of birds.

Site Moist, humus-rich, slightly acidic soil in partial to full shade.

Culture Amend soil with organic matter before planting. Once established, red baneberry will require little care, and it has no serious insect or disease problems. It spreads slowly to form showy clumps that seldom need division. Self-sown seedlings may appear.

Good Companions Flowering plants combine nicely with other taller woodland plants such as ferns, columbines, and wild geranium. Great lobelia, zigzag goldenrod, and large-leaved aster are good companions for the red berries in late summer.

Other Species *A. pachypoda* (white baneberry) is less prominent in Minnesota, found mainly along the eastern border. It is similar to red baneberry except for its berry color, which is white with a single black dot. It is also called doll's eyes because the berries resemble the china eyes once used in dolls. Use and culture are the same as for red baneberry. Zone 3.

Actaea pachypoda

Agastache foeniculum
Blue giant hyssop, anise hyssop
Zone 2

Agastache foeniculum

Native Habitat Dry, upland areas throughout most of Minnesota.

Height 2 to 4 feet

Description Blue giant hyssop has small, blue-purple flowers tightly packed into a 4- to 5-inch terminal spike in late summer. Leaves are 2 to 3 inches long, lance-shaped, pointed with serrated margins, and whitish underneath. Plant habit is upright, growing taller than it does wide.

Landscape Use Blue giant hyssop adapts well to perennial gardens or mixed borders, where it will offer great late-summer color and attract butterflies, hummingbirds, and bees. It is also well suited to prairie gardens. Grow it in herb gardens and harvest the anise-scented, edible leaves for teas, salads, and other drinks.

Site Full sun to partial shade in well-drained soil.

Culture Plants may self-sow, but they are easily weeded out. Staking may be required in perennial gardens.

Good Companions Contrast the blue-purple color with other late-summer, sun-loving perennials such as black-eyed Susan, oxeye, goldenrods, and coneflowers.

Cultivars 'Golden Jubilee' has gold-colored leaves and only grows to 20 inches. 'Alba' ('Album') and 'Snowspire' are white-flowered cultivars.

Allium cernuum
Nodding wild onion
Zone 3

Native Habitat Deciduous woodlands and open grasslands in a limited area along the Minnesota–Iowa border.
Height 12 to 24 inches
Description Nodding clusters of ¼-inch, purple-pink to white flowers hang from the down-turned tips of erect stems in June and July. Basal leaves are grasslike and grow up to 18 inches. The plant has a pleasant onion-like scent.
Landscape Use Nodding onion adds a touch of lavender to gardens in midsummer. The foliage is attractive all season, and the papery dried seed heads are decorative in autumn. Use it in open woodland gardens, rock gardens, and in mixed borders. It can be grown in containers.
Site Average to rich, well-drained soil in full sun to very light shade. Soil can be neutral to slightly alkaline.
Culture Nodding onion grows from bulbs that look like miniature versions of their cultivated relatives. It adapts readily to cultivation. Divide garden clumps every third year in early spring or as they go dormant to promote better flowering. Self-sown seedlings will appear. This plant is on the Minnesota threatened species list; make sure to grow only nursery-propagated plants.
Good Companions Plant nodding wild onion with wild petunia, purple coneflowers, prairie phlox, Culver's root, and obedient plant. Rattlesnake master and switch grass are good background plants.
Other Species *A. stellatum* (prairie wild onion) is native to open woodlands and prairies in all but the far southeastern corner of Minnesota. It has smaller, rose to pink flowers that are more upward facing when open. Plants grow 6 to 12 inches tall. It does well in cultivation when grown on a sunny exposure in a limy, rocky soil. Zone 3.

   A. tricoccum (wild leek) is native to rich, humus soils of cool, north-facing wooded slopes or limestone bluffs throughout most of Minnesota. It grows 6 to 16 inches tall in flower. The creamy white clusters of flowers appear on leafless stems in June and July after the two or three 8- to 10-inch, canoe-shaped leaves have withered away. These bright green leaves are showy when they first appear in early April on

Allium cernuum

Allium tricoccum

the forest floor when little else is up. It prefers moist, rich soil high in organic matter. The leaves need early season sunshine to generate food for flower production, but the plants grow in summer shade; a site under deciduous trees is ideal. Plant wild leek with wild ginger, which will fill in when the leaves die back. Zone 3.

Anemone canadensis
Canada anemone
Zone 2

Native Habitat Cool wet prairies and open woodlands in a variety of soils throughout Minnesota.
Height 12 to 24 inches
Description The long-stalked, snowy white, 2-inch flowers have numerous gold stamens. They appear in June. Leaves are deeply divided into three to seven lobes with toothed margins.
Landscape Use Canada anemone grows into large patches in moist prairies and open woods, and a large colony is quite striking in bloom. It makes a good groundcover beneath large trees and shrubs. It is good for filling in open or partially shaded areas. It tolerates moist soil and can be used in bog gardens.
Site Prefers moist, average to rich soil in full sun to light shade but tolerates a wide range of soils.
Culture Canada anemone can be aggressive in formal gardens and may need to be confined by edging strips buried in the soil. Plants are less aggressive in drier soil and in partial shade. Divide crowded plants in spring to encourage better blooming.
Good Companions Combine it with other native flowers such as butterfly weed, roses, columbines, black-eyed Susan, wild geranium, and bunch grasses in naturalistic plantings. It makes a nice yellow-white spring combination when planted with *Euphorbia polychroma* (cushion spurge) or *Doronicum* (leopard's bane) in mixed borders.
Other Species *A. quinquefolia* var. *quinquefolia* (wood anemone) is native to rich, moist soils in all but the southwestern corner of Minnesota. It grows 3 to 8 inches tall, forming large colonies with time. The solitary flowers are white above, pink to purple on their undersides, and bloom from late April to June. Leaves are palmately divided into three to five segments. It prefers partial

Anemone quinquefolia

Anemone canadensis

shade in moist, slightly acidic fertile soils in sheltered woodlands. It will eventually form a persistent carpet as it weaves its way among other plants. It goes dormant in early summer. Zone 3.

   A. cylindrica (long-headed thimbleweed) is native to prairies and dry savannas throughout Minnesota. It grows 2 feet tall with interesting greenish white flowers on long erect stalks that turn into cottony seed heads. It can be grown in moist to dry soil in full sun or light shade. Zone 3.

Anemonella [Thalictrum] thalictroides
Rue anemone
Zone 4; trial in zone 3

Anemonella thalictroides

Native Habitat Open, rocky woods and clearings in a variety of soils in southeastern Minnesota.

Height 4 to 8 inches

Description This delicate woodland plant has several long-stalked white, pink-tinged, or lavender-tinged flowers that radiate from a whorl of leaflets, each about 1 inch wide with three rounded lobes. They dance in the slightest breeze. It starts blooming in mid to late April, continues into June, and goes dormant by midsummer.

Landscape Use This cheerful harbinger of spring is attractive along path edges and as a carpet in woodland or shade gardens. It grows well in rocky terrain and can be planted in shady rock gardens. Keep in mind that it does disappear by midsummer.

Site Moist or dry, slightly acidic soil in filtered or partial sunlight.

Culture Keep plants blooming by watering during dry springs, since plants survive dry spells by going dormant early. Avoid competition from larger plants that may overtake this delicate plant. Plants are best left undisturbed.

Good Companions Create a beautiful spring carpet in wildflower gardens by growing rue anemones with delicate ferns and low groundcovers such as wild ginger and hepaticas. In shade gardens, combine it with early spring bulbs, primroses, violets, pulmonarias, and epimediums.

Cultivars 'Cameo' is a vigorous grower with soft pink double flowers. 'Green Dragon' has single green flowers. 'Green Hurricane' has interesting greenish double flowers held above wiry foliage. 'Schoaf's Double Pink' is a robust grower with deep rose-pink double flowers.

'Schoaf's Double Pink'

Antennaria plantaginifolia
Plantain-leaved pussytoes
Zone 3

Antennaria plantaginifolia

Native Habitat Open woodlands and woodland transition zones in a band from northwest to southeast across Minnesota.

Height 3 to 16 inches

Description This plant is mainly grown for its grayish, woolly leaves and stems. The leaves grow only a few inches tall. It spreads by runners. Dome-shaped clusters of fuzzy white flower heads appear late April into June on 12- to 16-inch stalks.

Landscape Use Pussytoes is a good groundcover for sunny areas near hot pavement or on banks, where it helps control erosion. The flat, silvery foliage is attractive all season and is effective in twilight gardens. It can also be used in rock gardens, between paving stones, and atop stone walls.

Site Thrives in dry, poor, well-drained soil in full sun but tolerates part shade.

Culture Requires little care once established. Divide in spring if you want more plants.

Good Companions Grow pussytoes with other drought-tolerant, sun-loving plants such as pasque flower and three-toothed cinquefoil. When planted with spring bulbs, it will fill in after the bulb foliage has died back.

Other Species *A. neglecta* (field pussytoes) is native to dry soils and prairies throughout Minnesota. It is similar to *A. plantaginifolia*, but the leaves are narrower and often more yellowish green. Plants are shorter, growing to 12 inches. Zone 3.

Aquilegia canadensis
Canada columbine, wild columbine
Zone 3

Aquilegia canadensis

Native Habitat Rocky woods and savannas in a variety of soils throughout Minnesota.

Height 12 to 24 inches

Description Columbine is a graceful, erect plant with nodding, upside-down, red and yellow flowers with five, upward-spurred petals dangling from the tips of branching stems. The grayish leaves are compound, divided into lobed leaflets grouped in threes. It blooms early May to early summer and the foliage is attractive all season.

Landscape Use Grow this charming plant anywhere, from woodland borders to prairies. Include it in rock gardens or scatter it around a garden pool. It attracts hummingbirds. Allow it to self-sow and form natural drifts in wild gardens.

Site Moist to dry, average, well-drained soil in full sun to almost full shade.

Culture Easily cultivated. Seedlings need moisture to become established, but deep rootstocks of mature plants survive dry spells. It will self-seed, but not to the point of becoming a pest. Old rootstocks do not transplant well. Leaf miners may attack the foliage, causing tan tunnels or blotches. Remove and destroy affected leaves as soon as you see them.

Good Companions Canada columbine's delicate nature combines nicely with a wide variety of plants. In native gardens, plant it with ferns, nodding wild onion, Canada anemone, prairie smoke, wild geranium, Virginia bluebells, and bird's-foot violet. In mixed borders, plant it near late tulips, hostas, iris, pulmonarias, peonies, and perennial geraniums.

Cultivars 'Corbett' is a pale yellow selection that combines beautifully with blue forget-me-nots or blue phlox in partial shade. 'Little Lanterns' is a diminutive selection growing only 8 to 10 inches tall.

Aralia racemosa
American spikenard
Zone 3

Aralia racemosa

Native Habitat Rich deciduous or mixed coniferous woods mainly in the eastern half of Minnesota.

Height 2 to 6 feet

Description Huge, twice pinnately divided, 2½-foot leaves have heart-shaped leaflets that are 4 to 6 inches long. Small terminal greenish flowers in May to July turn to showy, deep purple berries in late summer.

Landscape Use Spikenard is limited by its size, which relegates it to the accent or specimen category. Although it is herbaceous, think of it more as a shrub and use it to screen an area or as a foundation plant on the north or east side of a building. It becomes an understory plant in woodland gardens.

Site Rich, moist soil in partial to deep shade.

Culture Allow plenty of room for plants to spread. Established plants are difficult to move, so place plants in their permanent site if possible. Self-sown seedlings develop slowly and are usually not a problem.

Good Companions Combine American spikenard with other bold plants such as large ferns and hostas. Be careful about siting it next to smaller plants, which can be overtaken by this giant.

Other Species *A. nudicaulis* (wild sarsaparilla) has a slightly farther western distribution in Minnesota. It is a more diminutive but more aggressive relative, growing only 1 to 1½ feet tall. The umbrella-like, double-compound basal leaves shade three dome-shaped clusters of greenish white flowers growing on a leafless stem. Plants spread aggressively via underground rhizomes and may need to be contained by burying edging strips if you don't want them to become a groundcover. Zone 3.

Arisaema triphyllum
Jack-in-the-pulpit
Zone 3

Native Habitat Rich, moist woodlands throughout most of Minnesota.

Height 12 to 36 inches

Description Jack-in-the-pulpit has one or two long-stemmed compound leaves with three leaflets that form a canopy over the unusual flower. A ridged, hooded spathe with purplish brown streaks encloses an erect, brown spadix starting about mid-May. The bright red clusters of fruits are showy in late summer.

Landscape Use Jack-in-the-pulpit has an almost tropical look to it and catches the eye of many visitors. Use it in shady woodland gardens or mixed borders. The hooded spadix is interesting and intriguing to children. It will grow in heavily shaded, poorly drained soils.

Site Prefers a moist, rich soil in moderate to full shade.

Culture This easy-to-grow perennial adapts well to cultivation. Plants self-sow, but only to form a nice colony. When grown in drier conditions, it often goes dormant in summer.

Good Companions Jack-in-the-pulpit grows well with wild geranium, blue phlox, May apple, Canada columbine, and maidenhair fern.

Other Species *A. dracontium* (green dragon) is native to a strip along the southeastern border of Minnesota. Its long spadix is covered with tiny greenish yellow flowers that protrude well beyond the pointed spathe surrounding it. The solitary basal leaf is divided into five to fifteen leaflets. It grows 12 to 42 inches tall and blooms May to June. It prefers wet, rich soil in full sun to light shade. Showcase this unique plant along paths, near garden ponds, or against ledges. Zone 4.

Arisaema triphyllum

Asarum canadense
Wild ginger
Zone 3

Native Habitat Deciduous and some coniferous forests in nearly neutral or acidic soils mainly in the eastern half of Minnesota.

Height 6 to 8 inches

Description Wild ginger is a rhizomatous creeping plant with large, textured, heart-shaped leaves up to 8 inches wide. The leaves are up early in spring and they cover the interesting nodding, maroon flowers that appear in midspring.

Landscape Use Wild ginger is an excellent groundcover in shady areas. It forms an extensive carpet that is good for hiding empty spots left by spring ephemerals.

Site Prefers consistently moist, humus-rich soil in partial to full shade but tolerates drier, less acidic soils.

Culture Make sure new plants receive adequate water. Wild ginger is quite drought tolerant and carefree once established. It spreads quickly via creeping rhizomes but rarely becomes invasive. Divide plants in spring or as they go dormant in fall if you want to increase your numbers.

Good Companions This is a good plant to use among spring ephemerals such as trout lilies, false rue anemone, and prairie shooting star, which go dormant in summer. Combine it with foamflower (*Tiarella cordifolia*) in dappled shade to get a solid groundcover with widely differing leaf shapes and sizes.

Asarum canadense in flower

Asarum canadense

Asclepias tuberosa
Butterfly weed
Zone 3

Native Habitat Dry prairies and woodland edges and openings mainly in the southeastern quarter of Minnesota.

Height 1 to 3 feet

Description Butterfly weed has dense clumps of leafy stems topped with broad, flat clusters of fiery orange, red, or sometimes yellow flowers late spring to late summer.

Landscape Use Butterfly weed adds a splash of summer color to gardens. It is particularly striking when planted with complementary-colored blue and purple flowers. Plant it in perennial gardens, mixed borders, or prairie gardens. As the name implies, it attracts butterflies and their larvae, as well as hummingbirds, bees, and other insects.

Site Moist or dry soils in full sun or light shade. Mature plants can take full sun and dry soil.

Culture Set out young plants in their permanent locations, as the deep taproot makes plants difficult to move. Good drainage is essential; plants may rot in overly rich or damp soil. No major insect or disease problems, but plants may die over winter from root rot if the soil is too heavy. Plants are slow to emerge in spring, so cultivate carefully until new growth appears; you may want to mark the site each fall. Plants can get a little top heavy and may require gentle staking.

Good Companions Plant butterfly weed with other summer-blooming perennials such as purple salvias, white shasta daisy (*Leucanthemum × superbum*), and yellow daylilies for a showy display of contrasting shapes and colors. Good native companions include blazing stars, silky aster, leadplant, purple prairie clover, and wild bergamot.

Cultivars and Other Species 'Gay Butterflies' is a seed-grown strain of mixed yellow, red, and orange flowers.

Asclepias tuberosa

Asclepias incarnata 'Soulmate'

A. incarnata var. *incarnata* (swamp milkweed) is native to moist soil areas throughout Minnesota. It has flat, terminal clusters of pale rose to rose-purple flowers on 3-foot plants June to August. It grows best on constantly wet soils in full sun such as in bog gardens, but it adapts well to the conditions in sunny perennial borders if it receives supplemental water. It is a food source for several butterflies and their larvae. 'Ice Ballet' has creamy white flowers. 'Cinderella' and 'Soulmate' have fragrant, long-lasting, rose-pink flowers. Zone 3.

A. speciosa (showy milkweed) is native to prairies and savannas mainly in western Minnesota. It grows 1 to 3 feet tall and has clusters of rose-purple flowers. Plant it in prairie gardens where the damage caused by monarch butterfly caterpillars will not be a problem; it can become invasive in formal gardens. Zone 2.

A. sullivantii (Sullivant's milkweed, prairie milkweed) is native to prairies in south-central Minnesota. It is a rhizomatous species growing 2 to

3 feet tall with pink flowers best used in prairie gardens. It is on Minnesota's list of threatened plants; be sure to only purchase nursery-propagated plants. Zone 3.

A. syriaca (common milkweed) is native to average to moist soils throughout most of Minnesota. It has slightly drooping 2-inch clusters of mauve-pink, butterfly-attracting flowers on 24- to 60-inch downy stems. The undersides of leaves are covered with woolly gray hairs. It prefers moist but well-drained soil in full sun and is easy to grow in gardens but it can become invasive. Zone 3.

A. verticillata (whorled milkweed) is native to prairie soils in southern and western Minnesota. It has small, creamy white flowers in late summer. Plants spread from rhizomes to form large colonies. Zone 3.

A. viridiflora (green milkweed) is native to dry prairies and savannas in southern and western Minnesota. It has small greenish flowers in tight clusters. Grow it in average or sandy dry soil in full sun. Zone 3.

Aster novae-angliae
[Symphyotrichum novae-angliae]
New England aster
Zone 3

Native Habitat Mesic prairies in southwestern Minnesota.

Height 3 to 6 feet

Description New England aster flowers have violet or lavender petals surrounding yellow centers on heads 1 to 2 inches wide. Flowers are clustered at the ends of branches from late August into October. Mature plants have woody, fibrous root systems.

Landscape Use This fall bloomer brings purple shades to flower gardens. Use it in mixed borders, prairies, and even large rock gardens. It attracts butterflies and bees.

Site Moist, average soil in full sun to partial shade.

Culture New England aster likes consistent soil moisture. The tall stems can become top heavy when it's in bloom and need some type of support. Pinch stems back in late May to promote bushier plants. Divide plants in spring every third year to promote vigorous growth. Avoid too much nitrogen, which results in abundant foliage and floppy plants.

Good Companions Plant New England aster with goldenrods, grasses, oxeye, obedient plant, bottle gentian, showy tick trefoil, and boltonia for an outstanding fall show.

Cultivars and Other Species 'Andenken an Alma Potschke', usually sold simply as 'Alma Potschke', is covered in bright rose-pink flowers in fall. 'Purple Dome' is a dwarf cultivar (18 to 24 inches) with semi-double, deep purple flowers showy from late summer into fall. It is naturally dense and requires no pinching or staking.

You will find *Aster* species reclassified as *Symphyotrichum* in some references.

A. cordifolius (heart-leaved aster) is native to open woods and clearings mainly in southeastern Minnesota. The stems grow 3 to 4 feet tall with domes of white to sky-blue flowers in late summer. Give it a moist, humus-rich soil in light to partial shade. Zone 4.

A. ericoides (heath aster) is native to prairies and woodland edges throughout most of Minnesota. It has hundreds of white or pale blue flowers on 1- to 3-foot plants. Plant it in average to rich, moist soil in full sun. Zone 3.

A. laevis (smooth blue aster) is native to dry to moist prairies and woodland edges in all but the far northeastern corner of Minnesota. It has light blue flowers on 2- to 3-foot plants and is one of the last asters to bloom in autumn. Plant it in average to rich, moist but well-drained soil in full sun or light shade in woodland or prairie gardens where the floppy stems will be supported by nearby plants. Zone 2.

A. lateriflorus (calico aster, side-flowering aster) is native to open woods in poor or sandy soils mainly in the northern two-thirds of Minnesota. This 2- to 4-foot bushy plant has especially attractive foliage in addition to bearing hundreds of small white flowers. It forms clumps but is never invasive. Grow it in average to rich, well-drained soil in full sun or light shade. Zone 4; trial in zone 3.

A. macrophyllus (large-leaved aster) is native to woodland edges mainly in the northeastern half of Minnesota. It has flat-topped clusters of violet or lavender flowers with yellow to reddish centers on 1- to 4-foot stems August to September. Use it to create lush groundcovers in moist, humus-rich soil in sun or partial shade. It tolerates dry sites and deep shade and has good fall color. Zone 3.

A. oblongifolius (aromatic aster) is native to prairies in southwestern Minnesota. Its flowers are similar to *A. novae-angliae*, but it is more drought tolerant and blooms later. Zone 4.

A. oolentangiensis (sky-blue aster, azure aster) is found in dry to moist prairies and open woods in central and southeastern Minnesota. The light blue flowers are smaller and more open than smooth aster and it grows 2 to 5 feet tall. Give it well-drained soil in full sun. Zone 3.

A. sericeus (silky aster) is native to dry prairies and open woods in the southwestern half of Minnesota. It has lavender or pale blue flowers with yellow centers in loose clusters on many-branched, 12- to 24-inch stems. The silvery, silky leaves have smooth margins. It blooms late summer to early fall. Plant it in average soil in full sun to light shade in prairie gardens or mixed borders. It competes well with prairie grasses once established and is attractive to bees and butterflies. Zone 3.

A. umbellatus (flat-topped aster) is native to moist places throughout Minnesota. It has large, flattened clusters of small white flowers in late summer. It can grow up to 10 feet tall and may need support in some situations. It is a good companion for Joe-pye weed in moist soil in full or part shade. Zone 3.

Aster ericoides

Aster novae-angliae 'Purple Dome'

Baptisia lactea [B. alba, B. leucantha]
White wild indigo
Zone 4

Native Habitat Open woods, prairies, and savannas in moist or dry soils in southeastern Minnesota.

Height 3 to 5 feet

Description Wild white indigo has long, erect, pealike, white flowers in late spring. The compound leaves are an attractive bluish green color. Showy gray or brown seedpods rattle in the wind when they are ripe.

Landscape Use White wild indigo adapts well to cultivation. Use it in prairie gardens or in the middle to back of mixed borders, where the blue-green foliage provides a nice backdrop for smaller perennials. It forms a large clump, so leave plenty of space. The seedpods add interest in fall and winter and are often used in dried arrangements.

Site Well-drained soil in full sun or light shade.

Culture This long-lived perennial starts out slowly but eventually forms huge

Baptisia lactea

clumps that are difficult to transplant. Choose a site carefully, and space plants at least 3 feet apart to allow for growth. Plants rarely need dividing and resent disturbance. White wild indigo is drought tolerant once established. It is on the Minnesota special concern list and should be protected in the wild.

Good Companions Use white wild indigo as an accent plant with yarrows, artemisias, asters, and phloxes. Plant low, bushy plants such as geraniums around the base of clumps to hide the bare lower stalks. In prairies gardens, plant it with purple coneflowers, prairie phlox, butterfly weed, and prairie clovers.

Boltonia asteroides

Boltonia asteroides var. recognita
Boltonia
Zone 4

Native Habitat Wet prairies and marshes across the southern half of Minnesota.

Height 3 to 5 feet

Description Boltonia has small heads of white, aster-like flowers that have yellow centers. It starts blooming in late July and continues through September. It is an erect plant with narrow, gray-green leaves.

Landscape Use This summer bloomer provides the much-needed white color in late summer when yellows dominate perennial borders and prairie gardens.

Site Tolerates a wide variety of soils from wet to dry in full sun.

Culture Plants may need support to keep them from flopping over. They can be pinched back in late May to encourage compact growth. Overgrown clumps are easily divided every three to five years in spring.

Good Companions Plant boltonia with other late-summer perennials such as asters, goldenrods, fireweed, Joe-pye weed, and grasses.

Cultivars 'Snowbank' is more compact with sturdy stems smothered in white flowers in September. 'Pink Beauty' flowers a littler earlier than 'Snowbank' with attractive pale lavender-pink flowers.

Caltha palustris
Marsh marigold
Zone 2

Native Habitat Wet woods, swamps, and in the shallow water of ponds in northeastern Minnesota.

Height 12 to 24 inches

Description Marsh marigold has thick, branching stems bearing lustrous, dark green, heart-shaped or kidney-shaped leaves with lightly scalloped margins. The bright yellow flowers cover the plants for a long time, starting in mid to late April and continuing well into May.

Landscape Use Plant marsh marigold along stream banks, in bogs, or near water gardens in small clumps or large patches that can be seen from a distance.

Site Requires a wet soil high in humus in full sun to light shade. It is readily grown in shallow water or the wet soil of bog gardens.

Culture Amend soil with organic matter before planting. Marsh marigolds can be grown in containers in rich potting soil and covered with 2 inches of pea gravel before submerging in water. Plants go dormant about a month after flowering. Divide overgrown clumps as they go into dormancy.

Good Companions Pair with white-flowering wild calla (*Calla palustris*) to extend the season of color after marsh marigold goes dormant in midsummer. Other moisture-loving companions include pickerelweed (*Pontederia cordata*), arrowhead (*Sagittaria latifolia*), primroses, iris, and ferns.

Cultivars Var. *alba* has pure white flowers. 'Flore Pleno' has fully double, long-lasting flowers.

Caltha palustris

Campanula rotundifolia
Harebell, bluebell
Zone 2

Native Habitat Moist to dry prairies and savannas in southeastern Minnesota.

Height 6 to 18 inches

Description Harebell has violet-blue, bell-shaped, nodding flowers that appear on slender stalks at ends of branched stems. It blooms June to September.

Landscape Use Plant harebell where the delicate flower won't be overpowered by nearby plants. It does well in rock walls, in rock gardens, and between pavers on a terrace, where it will bloom all summer.

Site Needs a well-drained soil in full sun.

Culture Add sand and organic matter to improve soil drainage if necessary. Avoid overly rich soils, which can encourage vigorous growth on nearby plants that can overtake harebell. In the right setting, the underground stems of harebell will spread.

Good Companions Plant harebell with other diminutive dry-soil plants such as creeping thymes, pussytoes, alliums, columbines, prairie smoke, and beardtongues.

Other Species *C. americana* (tall bellflower, American bellflower) is native to prairies and open woodlands in the southeastern quarter of Minnesota. It is an annual or biennial that will self-sow. The light blue flowers grow in clusters atop 4- to 6-foot stems. The blooms progress up the stem and appear spring to summer. Plant it in moist, average soil in full sun to partial shade. It is sometimes listed as *Campanulastrum americana*. Zone 3.

Campanula rotundifolia

Caulophyllum thalictroides

Caulophyllum thalictroides
Blue cohosh
Zone 3

Native Habitat Rich, damp woods and bottomlands throughout most of Minnesota.

Height 1 to 3 feet

Description This handsome, erect plant has a thrice-compound leaf with an attractive sea green color. Stems have a purplish tint when they emerge in spring. The yellowish green or purplish flowers in May are rather small and are followed by showy berrylike, midnight blue fruits borne in a loose cluster.

Landscape Use Use blue cohosh in shade or woodland gardens. Be sure to place it where you can enjoy the striking early spring purple stems and the bright blue fruits in late summer.

Site Prefers a moist, well-drained soil with abundant organic matter in partial to full shade.

Culture Plants form multistemmed clumps in time. Divide in early fall and cut apart crowns, leaving at least one eye per division.

Good Companions Plant blue cohosh with early spring ephemerals, which will set off the purple emerging stems. It will fill in when Dutchman's breeches, spring beauty, false rue anemone, trout lilies, and other ephemerals go dormant. It grows naturally with two-leaved miterwort, which also makes a good garden companion. The fall berries combine nicely with zigzag goldenrod and white snakeroot.

Caulophyllum thalictroides fruits

Chelone glabra
White turtlehead
Zone 3

Native Habitat Wet prairies and open woods mainly in the eastern half of Minnesota.

Height 1 to 3 feet

Description White turtlehead has terminal clusters of white, inflated, arching, two-lipped flowers that are often tinged with pink or lavender. They appear in late summer and early fall and are followed by attractive dried seed heads. Plants have an upright to slightly vase-shape form and dark green leaves.

Landscape Use Plant white turtlehead in flower borders, wet prairies, and alongside water features. The flowers attract bees and hummingbirds.

Site Evenly moist to wet, rich soil in full sun to light shade.

Culture Mulch plants well to conserve soil moisture. Plants will tolerate brief dry spells once they are established. Plants spread by rhizomes but do not become invasive.

Good Companions In flower borders, white turtlehead combines nicely with asters, prairie phlox, goldenrods, Joe-pye weed, and grasses. Near ponds, grow it with iris, cardinal flower, ferns, and obedient plant.

Other Species *C. obliqua* (purple turtlehead) is native to wet woods and marshes in southern Minnesota. It grows 1 to 3 feet tall with rich pink to rose-red flowers on upright to arching stems. Grow it in moist to wet, humus-rich soil in full sun to part shade in the same places as white turtlehead. Zone 4.

Chelone glabra

Claytonia virginica
Virginia spring beauty
Zone 3

Native Habitat Moist woods and by streams along the eastern side of Minnesota.

Height 6 to 8 inches

Description This dainty spring ephemeral has charming white or white and pink candy-striped flowers in loose clusters in April and early May. The narrow leaves appear in basal clumps 4 to 6 inches tall.

Landscape Use Spring beauty makes a beautiful but short-lived spring groundcover in woodland gardens. Plant it in masses to intermingle with other early wildflowers and bulbs, or showcase a group of three to five plants at the base of a tree. It disappears soon after flowering, so interplant it with ferns and other summer flowers. It can be naturalized in lawns. Plant a few plants, let them reseed, and do not mow the grass until they go dormant.

Site Prefers moist, rich soil in light shade or spring sun.

Culture Spring beauty grows from a small, tuberlike corm that is easily divided after flowering to increase the population. It also self-sows. Be careful not to disturb the dormant clumps in summer. Plants can take considerable drought in summer, but need consistent soil moisture in spring and fall.

Good Companions Plant spring beauties with other ephemerals such as trout lilies, cut-leaved toothwort, and Virginia bluebells, along with persistent plantings such as wild ginger, ferns, and baneberries. Mix it with spring bulbs such as snowdrops, species crocus, and glory-of-the-snow under flowering shrubs and trees.

Other Species *C. caroliniana* (Carolina spring beauty) is native on south- and east-facing slopes in mesic, old-growth northern hardwood forests along Lake Superior. It has shorter, broader leaf blades and fewer leaves and flowers, which are white to pale pink. It spreads slower than Virginia spring beauty but will eventually form a dense carpet. It is on Minnesota's special concern list. Zone 3.

Claytonia virginica

Clintonia borealis
Bluebead lily
Zone 2

Native Habitat Rich, acidic coniferous woods in the northeastern quarter of Minnesota.

Height 6 to 15 inches

Description The steel-blue berries that adorn this plant in late summer are more distinctive than its loose cluster of greenish yellow, nodding, bell-shaped flowers that appear late May to mid-June. The shiny, oblong basal leaves are 5 to 8 inches long and resemble flattened tulip leaves.

Landscape Use Bluebead lily's bold texture brings interest to shade or woodland gardens. Plant it in masses as groundcover or mix scattered clumps with other finer-textured plants. Children should be discouraged from eating the mildly poisonous berries.

Site Moist, high-humus soil in partial to deep shade.

Culture Bluebead lily is somewhat difficult to grow in cultivated settings. It requires cool, shady, moist sites and prefers temperatures below 75 degrees.

Good Companions Grow bluebead lily with other northern wildflowers such as twinflower (*Linnaea borealis*), starflower (*Trientalis borealis*), trilliums, and ferns.

Clintonia borealis

Coreopsis palmata
Bird's-foot coreopsis, stiff tickseed
Zone 4

Coreopsis palmata

Native Habitat Dry prairies in central and southeastern Minnesota.

Height 2 to 3 feet

Description Bright yellow, daisylike flowers top fine-textured plants June through August.

Landscape Use Bird's-foot coreopsis is a cheery, old-fashioned plant that attracts butterflies. It spreads rapidly by underground rhizomes to form a dense mat and is a good choice for stabilizing dry, sunny slopes, but it is probably too aggressive for most landscape situations.

Site Moist or dry soils in full sun or light shade. Mature plants can take full sun and dry soil.

Culture Bird's-foot coreopsis is a tough, low-maintenance garden plant. Mature plants tolerate summer drought. Plants may flop over if the soil is too rich. Regular deadheading will prolong flowering.

Good Companions Plant bird's-foot coreopsis with other summer-blooming prairie plants such as butterfly weed, wild bergamot, flowering spurge, and purple prairie clover.

Cornus canadensis
Bunchberry
Zone 1

Cornus canadensis

Cornus canadensis fruits

Native Habitat Moist, acidic woodlands and bogs in the northeastern half of Minnesota.

Height 4 to 8 inches

Description Bunchberry has a whorl of satiny, oval evergreen leaves 2 to 3 inches long. What appears to be a single flower atop the stem is actually a cluster of small, greenish yellow flowers surrounded by four creamy white, petal-like bracts late May into June. Clusters of bright orange-red berries appear in summer. This subshrub grows from creeping, wiry stems to form extensive colonies.

Landscape Use Bunchberry makes an effective groundcover under acid-loving plants such as azaleas. The showy flowers and berries brighten shady recesses in wild gardens. Plant it in colonies for best effect.

Site Requires moist, acidic, high-humus soil in light shade.

Culture If necessary, acidify soil to a pH of 4 to 5 before planting and work in ample organic matter. Conifer needles are a good acidic mulch. Soil should be constantly moist but not waterlogged. It will not tolerate summer heat or drought. It is difficult to keep garden plants under less-than-optimum conditions.

Good Companions Plant bunchberry with other northern wildflowers and shrubs such as wintergreen, starflower, Canada mayflower, blue-bead lily, low-bush blueberry, leather leaf, and ferns.

Cypripedium calceolus var. pubescens
Large yellow lady's-slipper
Zone 3

Native Habitat Bogs or moist woods, mostly along north-facing slopes in northeastern Minnesota.

Height 12 to 24 inches

Description Yellow lady's-slipper has leafy stems bearing one or two flowers, each consisting of a pale yellow pouch 2 inches long flanked by two petals and two greenish brown, twisted sepals. Leaves are 6 to 8 inches long, with parallel veins and smooth margins. It blooms in May.

Landscape Use This showy plant should be used as a specimen or massed in moist shade or woodland gardens in a spot where it can be enjoyed while in bloom. It is long lived and resents transplanting, so choose a site carefully.

Site Wet, high-humus soil in light shade.

Culture Yellow lady's-slipper is the easiest native orchid to grow. Prepare the soil well before planting it 12 inches apart in groups of three or randomly. Maintain a 1- to 2-inch mulch layer to keep the soil moist. It will form large clumps that can be carefully divided in spring. Exercise caution when obtaining plants: all lady's-slippers are difficult to propagate, and many commercially available plants were probably collected in the wild.

Good Companions Grow this showy plant with background plants that will complement rather than compete with it, such as low-growing ferns and small-flowered wildflowers like wood anemone.

Varieties Var. *parviflorum* (small yellow lady's-slipper) is native to bogs and low woods in northeastern Minnesota. It grows 10 to 14 inches tall with two to four 1½-inch flowers per spike. The lip is a darker yellow, and it blooms a week or two later. Zone 3.

Cypripedium calceolus var. pubescens

Dalea purpurea var. purpurea
[Petalostemum purpureum]
Purple prairie clover
Zone 3

Native Habitat Dry or moist prairies, savannas, and open woods mainly in the southwestern half of Minnesota.

Height 2 to 3 feet

Description Purple prairie clover has unique, densely packed, ½- to 2-inch, rose-purple to crimson flowers in June and July. The flowers bloom in a ring around the flower head, starting at the bottom and working up to the top. The foliage is pinnate and fine-textured. Seed heads are attractive in winter.

Landscape Use This prairie denizen is attractive to bees and butterflies. Plant it in groups of three to five in perennial beds, prairie gardens, and xeriscaped beds. A single plant can be used as an accent in rock gardens.

Site Moist to dry soils in full sun.

Culture Purple prairie clover is a tough, low-maintenance garden plant. Mature plants tolerate summer drought, and clumps seldom need dividing. It fixes soil nitrogen so it doesn't need additional fertilizer. You may need to use an inoculant to help plants become established in some soils. Allow seed heads to remain for winter interest.

Good Companions Plant purple prairie clover with other summer-blooming flowers such as wild bergamot, leadplant, butterfly weed, coreopsis, mountain mint, ornamental onions, and smaller grasses like little bluestem and prairie dropseed.

Other Species D. candida (white prairie clover) is native to dry prairies and savannas in southwestern Minnesota. It grows 3 to 4 feet tall and has less-showy white flowers. It is on Minnesota's special concern list, so only grow nursery-propagated plants. Zone 3.

Dalea purpurea

Dalea candida

89

Dentaria laciniata
[*Cardamine concatenata*]
Cut-leaved toothwort
Zone 3

Dentaria laciniata

Native Habitat Moist deciduous woods, bottomlands, and streamsides mainly in southeastern Minnesota.

Height 4 to 12 inches

Description Drooping, white to pink flowers appear in loose terminal clusters April to June. Midway up the stem is a whorl of three deeply divided, toothed leaves.

Landscape Use This beautiful plant will eventually form large colonies that weave through other plants in woodland gardens, but it is not as vigorous in spreading as some other ephemerals.

Site Moist, high-humus soil. It tolerates fairly dense shade and drier soils in summer but needs some direct sun in spring.

Culture Weed out aggressive neighbors before planting and as they appear. Plants go dormant in early summer but may appear again in fall.

Good Companions Plant cut-leaved toothwort with other spring wildflowers such as Virginia bluebells, Dutchman's breeches, false rue anemone, and spring beauties. Be sure to include some persistent plants such as ferns and wild ginger to fill the void when these ephemerals go dormant.

Desmodium canadense
Canada tick trefoil
Zone 3

Desmodium canadense

Native Habitat Tall-grass prairies and woodland openings mainly in the southwestern half of Minnesota.

Height 2 to 4 feet

Description This plant has elongated terminal clusters of bright purple, pea-like flowers on well-branched, bushy plants with trefoil-shaped leaves. It blooms July to early September.

Landscape Use Plant Canada tick trefoil in prairie gardens or in the back of perennial borders, keeping in mind its somewhat aggressive nature. The seeds are eaten by a variety of birds and other wildlife.

Site Does well in clay soil, but will grow in any moderately rich, well-drained or slightly damp soil in sun to light shade.

Culture Canada tick trefoil may be difficult to establish because of its deep taproot. It is drought tolerant once established. A member of the legume family, it improves the soil by adding nitrogen. Plant it with a commercial soil inoculant to provide the root nodules with the necessary nitrogen-fixing bacteria. It can become large and aggressive, so plant it where there is growing room.

Good Companions Combine Canada tick trefoil with other summer-blooming prairie natives such as leadplant, bird's-foot coreopsis, butterfly weed, Culver's root, and bottle gentian.

Dicentra cucullaria
Dutchman's breeches
Zone 3

Native Habitat Moist, often rocky deciduous woods in all but northwestern Minnesota.

Height 6 to 12 inches

Description Dutchman's breeches has delicate, fernlike, blue-green basal foliage and unique blooms that resemble white pantaloons hanging upside down from arched stems. Flowers are white with traces of yellow. Blooms first appear in mid to late April and last for about two weeks. Plants go dormant mid to late May.

Landscape Use Use this dainty ephemeral in woodland or shade gardens. In the wild, it is often found growing tucked into root flares of large trees; it tolerates this shallow soil found around mature tree roots because plants go dormant before it becomes too dry.

Site Needs spring sun and moist soil to bloom well, but can grow in shade and dry soil once plants are dormant.

Culture Grows from tuberous rhizomes. You may need to cover dormant plants with hardware cloth to discourage rodents from stealing the tubers. Colonies often form a dense carpet of foliage punctuated by clusters of flowers. Plants disappear soon after flowering. Self-sown seedlings will appear.

Good Companions Plant Dutchman's breeches with other spring ephemerals as well as plants that will fill in blank spots such as pachysandra, ferns, and wild ginger.

Other Species *D. canadensis* (squirrel corn) occurs naturally in moist deciduous forests in the southeastern corner of Minnesota. It has white, heart-shaped flowers ½ to ¾ inch long that dangle from 6- to 12-inch stems above compound, finely cut foliage that resembles Dutchman's breeches. It blooms mid-April to June, about a week later than Dutchman's breeches. Plant it in moist, high-humus soil in direct or filtered spring sun. It is on Minnesota's special concern list; be sure to purchase only nursery-propagated plants. Zone 4.

Dicentra cucullaria

Dodecatheon meadia var. *meadia*
Prairie shooting star
Zone 4

Native Habitat Fertile, moist woods and mesic calcareous prairies in the far southeastern corner of Minnesota.

Height 10 to 20 inches

Description The flowers have delicate, white to pink, strongly reflexed petals surrounding a yellow and red conelike center, giving the appearance of a shooting star. The lush green basal foliage has reddish tints at the base. Plants bloom for a long time, late April to June.

Landscape Use Prairie shooting star has somewhat of a split personality, being both a prairie and a woodland flower. A spring bloomer, it's at its best while most other prairie flowers are just appearing. Showcase this interesting plant along walkways, in rock gardens, or in perennial borders. It can also be grown in partial shade areas in woodland or shade gardens. Plant it in groups of at least three, and do not crowd it with other species.

Site Any good soil, in sun or light shade, but does best in rich garden soil. Requires direct sun in spring but shade is okay in summer.

Culture The ideal soil pH is 6 to 7; add limestone before planting if necessary. Plants need moisture while blooming, but are drought tolerant after that. Plants develop slowly and take several years to bloom from seeds. Fall transplanting is recommended, but planting in early spring is also fine. The basal rosette of leaves disappears after flowering, by early August, to make way for summer- and fall-blooming flowers. Prairie shooting star is on Minnesota's endangered species list; be sure to purchase only nursery-propagated plants.

Dodecatheon meadia

Good Companions In shadier spots, plant prairie shooting star with Canada columbine, hostas, ferns, and primroses. In sunnier sites and prairie gardens, it combines nicely with golden alexanders, prairie smoke, prairie phlox, wild geranium, and violets.

Cultivars 'Album' is a white selection. 'Goliath' is several inches taller than the species and has large, lavender-rose blooms.

Echinacea angustifolia
[E. pallida var. angustifolia]
Narrow-leaved purple coneflower
Zone 3

Echinacea angustifolia

Native Habitat Open woods, savannas, and prairies mainly in southwestern and western Minnesota.

Height 1 to 3 feet

Description Narrow-leaved purple coneflower has stout, hairy, nearly leafless stems topped with heads of pale rose-pink petals surrounding a dark purple-brown central disc. It blooms in summer, usually late June into August.

Landscape Use *E. angustifolia* is not as showy as it its more popular cousin, *E. purpurea* (purple coneflower), which is native farther south and east, but it is still a good addition to prairie gardens and perennial borders. The flowers are attractive to butterflies, and goldfinches enjoy the seeds.

Site Average to rich soil in full sun or light shade.

Culture Divide only when you have to, since divisions usually don't produce as many flowers. Drought tolerant once established. Self-sown seedlings will appear.

Good Companions In prairie gardens, plant it in clumps near other summer-blooming prairie flowers such as black-eyed Susan, wild bergamot, bird's-foot coreopsis, compass plant, blazing stars, butterfly weed, and tall bellflower, as well as native grasses. In perennial borders, combine it with phloxes, yarrows, delphiniums, and baby's breath.

Eryngium yuccifolium
Rattlesnake master
Zone 4

Eryngium yuccifolium

Native Habitat Open woodlands and mesic prairies in southern Minnesota.

Height 2 to 3 feet

Description Rattlesnake master has clusters of small, greenish white, fragrant flowers in summer. They are tightly packed among pointed bracts to form globular flower heads about ¾ inch wide in branched clusters atop the stems. The blue-green leaves are large and narrow with pointed teeth and clasping bases. Seed heads turn brown and remain on the plant for winter interest.

Landscape Use Rattlesnake master works well in sunny borders or prairie gardens. The distinctive yuccalike leaves add interest when plants are not in flower. It can be used in a formal setting as an accent plant.

Site Prefers average to rich, moist but well-drained soil in full sun. Plants will grow in gravel and sand in full sun.

Culture Add sand or gravel to improve soil drainage if necessary. Young leaves may need protection from rabbits and deer in spring. The deep taproot resents disturbance, so set out young plants in their permanent locations. Plants seldom need division. Rattlesnake master is on Minnesota's special concern list, so only buy nursery-propagated plants.

Good Companions Combine bold rattlesnake master with airy clusters of flowering spurge, baby's breath (*Gypsophila paniculata*), and sea lavender (*Limonium latifolium*). The flowers look nice with blazing stars, Culver's root, Michigan lily, black-eyed Susan, coneflowers, goldenrods, Culver's root, autumn sneezeweed, asters, and grasses.

Erythronium americanum
Yellow trout lily
Zone 3

Native Habitat Moist, rich woods often along streams mainly in the southeastern quarter of Minnesota.

Height 4 to 10 inches

Description Yellow trout lily has a pair of maroon-mottled green leaves 3 to 8 inches long, narrowly oval with smooth margins. The leaves clasp the base of a stem bearing a single yellow flower with three petals and three sepals, all reflexed. Plants grow from bulblike corms and bloom in April and early May. They form large mats of foliage on the forest floor but don't always produce a lot of flowers.

Landscape Use This ephemeral is one of the earliest wildflowers to bloom. Plant it in woodland gardens, where the interesting foliage makes an attractive spring groundcover.

Site Likes lots of spring sunlight in a moist, rich soil. Once plants go dormant, they can be in shade.

Culture Add generous amounts of humus to the soil before planting. Plants may take five or six years before they flower. Plants self-sow and spread by stolons, eventually forming large drifts. Plants go dormant right after flowering; be careful not to dig into clumps in summer or fall.

Good Companions Interplant yellow trout lily with minor spring bulbs such as purple *Scilla siberica* (Siberian squill) and other spring ephemerals such as Dutchman's breeches, spring beauty, and bloodroot. Include ferns and foliage plants such as wild ginger, pulmonarias, and hostas to fill in after the plants go dormant.

Other Species *E. albidum* var. *albidum* (white trout lily) is found in low, deciduous woods mainly in southeastern Minnesota. The petals are white or dull pale violet-tinged outside and yellow-tinged inside and the leaves are a slightly lighter greenish gray. Petals are strongly recurved. It flowers more profusely in cultivation than yellow trout lily. Grow it in moist, rich, neutral soil in woodland or shade gardens. Zone 3.

E. propullans (dwarf trout lily) is unique to Minnesota, occurring on wooded, north-facing slopes near streambeds only in Rice and Goodhue counties in east-central Minnesota. This rare native is on both the state and federal endangered species lists. Flowers are smaller than *E. albidum* and are pinkish rather than white. Fruits are also smaller and are nodding rather than erect when mature. Attempts to propagate this species have been unsuccessful, so it is not practical for garden use. It is included here only because of its endemic nature.

Erythronium americanum

Erythronium albidum

Eupatorium purpureum
Sweet Joe-pye weed
Zone 3

Native Habitat Wet slopes, low prairies, and woodland edges mainly along the eastern edge of Minnesota.

Height 3 to 6 feet

Description Sweet Joe-pye weed has mounded or domed clusters of pale rose or smoky lavender, sweet-scented flowers. The leaves appear in whorls around the stems, which are often a deep purple color. Seed heads are attractive in winter.

Landscape Use Use sweet Joe-pye weed in prairie gardens, perennial beds, and mixed borders. It thrives at the edges of water gardens or in bog gardens. Butterflies are attracted to the flowers.

Site Prefers moist, average to rich soil in full sun or light shade but tolerates drier conditions.

Culture Good garden plant; easy to grow once established. Takes at least two seasons for new plants to reach full size. Will perform well in dry situations, but best growth is in moist soil. Plants are late to emerge in spring, but grow quickly after they are up. Several insects eat the leaves, which can leave plants looking a little rough in midsummer, but the damage is not life threatening. Divide oversized clumps in spring or fall. Reseeds. Plants too close together may suffer from powdery mildew.

Good Companions Many plants look nice with a backdrop of sweet Joe-pye weed, including autumn sneezeweed, boltonia, coneflowers, black-eyed Susans, and daisies. In less-formal situations, plant it with asters, wild bergamot, goldenrods, and grasses.

Other Species and Cultivars

E. maculatum (spotted Joe-pye weed) is native to wet prairies and other moist-soil areas throughout Minnesota. It has flat-topped, 4- to 5-inch clusters of feathery, rose-purple flower heads on top of purple or purple-spotted stems 24 to 72 inches tall. It blooms late July to September in wet, average soil in full sun. Use it in wet prairies or naturalized plantings. 'Gateway' grows 6 feet tall or more and has pale mauve flowers in large clusters. It has sturdier, richly colored stems. 'Atropurpureum' is more compact, up to 5 or 6 feet, with deep purple stems, dark leaves, and purple, sweet-scented flowers. Both cultivars are suitable for garden use. Zone 3.

E. perfoliatum var. *perfoliatum* (common boneset) is native to moist prairies and swamp margins scattered throughout the state. It grows 2 to 4 feet tall with short, hairy stems supporting flat-topped clusters of off-white flower heads in mid to late summer. Leaves have a wrinkled texture and bases are fused so the stem appears to be pierced by a single long leaf. Requires consistently moist or wet soil in full sun. Use it in wet prairies or bog gardens, where it attracts bees and butterflies. Zone 3.

E. rugosum (white snakeroot) is native in open woods across the southern half of Minnesota. It grows 3 to 4 feet tall with white flowers in small clusters in late summer. The seed heads are showy. It can tolerate drier soil in shaded woodland gardens, but doesn't like full shade. Some references have started classifying it as *Ageratina altissima*. Zone 3. 'Chocolate' has dark chocolate brown leaves and shiny purple stems, which are a great contrast with the clusters of domed white flowers in fall. It grows 2 to 4 feet tall. The silver seed heads offer fall and winter interest. Grow it with golden-leaved hostas to set off color contrast. Zone 4.

Eupatorium purpureum

Eupatorium perfoliatum

Eupatorium rugosum

Euphorbia corollata var. corollata
Flowering spurge
Zone 3

Native Habitat Open woods, savannas, and prairies in southeastern Minnesota.

Height 2 to 4 feet

Description Flowering spurge has clusters of small white flowers with pure white bracts that give it a look of a sturdy baby's breath. It blooms June into September. The creeping slender stems have sparse pale green leaves, and plants bleed milky sap when picked or damaged. The foliage turns an attractive orange-gold color in fall.

Landscape Use The airy flower heads are easy to use in gardens, but its aggressive nature makes flowering spurge best suited to large sunny borders or naturalized plantings, where it weaves and helps tie plantings together. Because it grows well on dry, clay hillsides and roadsides where few other plants thrive, it's often used for erosion control and restoration projects.

Site Poor to average, well-drained soil in full sun or partial shade.

Culture This long-lived perennial tolerates drought and poor soil and needs little care once established. Plants grow from creeping rhizomes to form dense stands. Mature plants prefer to be left alone, but clumps can be divided as needed to control spread or for propagating. The white sap that oozes from cut stems can be irritating to skin.

Good Companions Good prairie companions include butterfly weed, blazing stars, coneflowers, wild bergamot, oxeye, Joe-pye weeds, goldenrods, and asters. In dry-soil gardens, plant it with sea hollies (*Eryngium*), sages, and ornamental grasses.

Other Species *E. marginata* (snow-on-the-mountain) is found in prairies scattered across southern Minnesota. It is an annual with dense clusters of showy bracts outlined in white. Upper leaves

Euphorbia corollata

are variegated, ranging from mostly white at the top of the plant to all green on the lower stem. It will reseed. Use it in prairie gardens or on sites with dry, poor soil in full sun. Zone 3.

Gaillardia aristata
Blanket flower
Zone 2

Native Habitat Dry prairies in far northwestern Minnesota.

Height 18 to 24 inches

Description This perennial has showy, 2- to 4-inch, yellow petals with a narrow band of burgundy at the base where they join the dark red disk flowers. Leaves are hairy and lobed or egg shaped. It blooms June to August.

Landscape Use Blanket flower is a good addition to perennial borders, where it attracts bees and butterflies. It can also be used as a cut flower or container plant.

Site Grows best in average to poor soil in full sun.

Culture Blanket flower is easy to grow in most garden settings. If the soil is heavy, add generous amounts of sand and organic matter before planting. Remove old flower heads early in the season to prolong bloom. It is on Minnesota's special concern list, so only buy nursery-propagated plants.

Good Companions Set off blanket flower's showy bicolor blooms by planting it with single-colored summer bloomers such as Mexican hat, snow-on-the-mountain, coreopsis, butterfly weed, and wild bergamot.

Cultivars *G. × grandiflora* is a showy cross between *G. aristata* and *G. pulchella*, another North American native found in more-southern locations. The resulting floriferous hybrids have orange and yellow flowers, often with dark red bands or eyes. Plants are showy but short-lived. They adapt well to container culture. 'Burgundy' has large wine-red flowers. 'Dazzler' has golden-yellow flowers with maroon centers. 'Goblin' is a 12-inch plant with red-centered flowers with yellow edges. Zone 4.

Gaillardia × grandiflora 'Goblin'

Gaultheria procumbens
Wintergreen
Zone 3

Native Habitat Dry or moist woods, outcroppings, and forested bogs in northeastern Minnesota.

Height 2 to 6 inches

Description Wintergreen has white, nodding, bell-shaped flowers that grow singly or in groups of two or three from the leaf axils of the creeping woody stem. The flowers turn to showy bright red berries. Leaves are dark green, shiny, and evergreen. Flowers bloom April to May and the red fruits persist among evergreen leaves all winter into spring.

Landscape Use This low-growing plant has trailing woody stems that form broad, irregular colonies. It is a good groundcover under acid-loving shrubs such as azaleas. The bright red berries and evergreen foliage offer win-

Gaultheria procumbens

ter interest and can even brighten early spring gardens if birds do not eat the fruits.

Site Tolerates a wide range of conditions but requires acidic soil with moderate humus content in partial shade.

Culture Acidify soil before planting, if necessary. Plants are slow to establish. Water well the first year. Divide rhizomes in spring to increase the population.

Good Companions Include wintergreen in northern coniferous garden habitats, where it looks nice with hepaticas, Canada mayflower, bunchberry, and wild ginger.

Other Species *G. hispidula* (creeping snowberry) grows in bogs and damp coniferous woods and on mossy logs in northeastern Minnesota. It has intermittent, tiny, white flowers along short, trailing stems that form a delicate mat. Stems grow 3 to 12 inches long. It blooms May to June and has snow-white berries later in the season. It needs a moist, rich, acidic soil in full shade and can be used as groundcover for moist, shady sites. Zone 3.

Gentiana andrewsii
Bottle gentian
Zone 3

Native Habitat Moist, open woods, wet prairies, and marshes throughout most of Minnesota.

Height 1 to 3 feet

Description Bottle gentian has deep blue, 1-inch, bottle-shaped flowers crowded into a terminal cluster, blooming August to October. Other flowers may cluster in the axils of the opposite leaves below. Plants are erect to sprawling with glossy, oval, 4-inch leaves. Since the blossoms never open, pollinating insects must burrow through the petals.

Landscape Use Bottle gentian's rich indigo blue color is a welcome sight in autumn prairie gardens when most plants are orange or yellow. It adapts well to the rich soil of perennial borders and can be grown in wet meadows and along ponds.

Site Prefers a sandy loam high in organic matter that stays moist throughout the growing season. Tolerates sun or partial shade; afternoon shade from summer sun is beneficial.

Culture Bottle gentian is easy to grow and long lived, requiring little care once established. Make sure the soil is high in organic matter, and mulch plants to conserve soil moisture. Plants seldom need dividing.

Good Companions Bottle gentian is a great companion for any of the fall prairie plants, including New England aster, obedient plant, and blazing stars. In perennial borders, use it with other moisture-loving plants such as turtlehead, ferns, and lobelias.

Other Species *G. flavida* [*G. alba*] (yellowish gentian) is native to dry to medium prairies and open woods in southeastern Minnesota. It has creamy white flowers over semi-glossy, medium-green to yellow-green leaves. Plants bloom in late summer. Plants are more upright than *G. andrewsii*, reaching a height of 2 feet or more. Use it in moist prairie plantings or in mixed borders. It is a long-lived plant that grows into a good-sized clump with time. Zone 3.

Gentianopsis crinita (greater fringed gentian) is native to moist or wet open sites widely scattered in Minnesota. This close relative (it is some-

Gentiana andrewsii

Gentiana flavida

times listed as *Gentiana crinita*) has stunning, upward-facing, fringed flowers in satiny blue violet. It grows 1 to 3 feet tall and is one of the few gentians to have open flowers. This biennial doesn't reliably reseed and must be sown each year. It is a challenge to grow, but is rewarding when successful. Zone 4.

Geranium maculatum
Wild geranium
Zone 3

Native Habitat Open deciduous woods, bottomlands, prairies, and savannas mainly in the southeastern quarter of Minnesota.
Height 12 to 24 inches
Description Wild geranium has loose clusters of five-petaled, inch-wide flowers rising above pairs of grayish green leaves, each with three to five distinct palmate lobes and coarse teeth. Flowers appear early May into June, ranging from pale to deep magenta-pink to light purple. The fruits that appear after flowering resemble a crane's bill. Foliage turns a lovely red color in fall.
Landscape Use Wild geranium is easily adapted to culture. Plant it in drifts in woodland gardens. In perennial borders, it can grow in part sun.
Site Prefers moist, rich soils in partial sun to light shade, but tolerates drier conditions.
Culture Wild geranium is easy to grow and transplant. Divide the slow-creeping rhizomes in early spring or early fall. Shelter plants from strong winds.
Good Companions Combine wild geraniums with other woodland denizens such as golden alexanders, blue phlox, Canada columbine, and ferns.
Cultivars 'Album' and 'Hazel Gallagher' are white-flowered cultivars.

Geranium maculatum

Geum triflorum
Prairie smoke
Zone 2

Native Habitat Dry to moist gravel or black-soil prairies and savannas across the southwestern half of Minnesota.
Height 6 to 16 inches
Description Prairie smoke has pink- or rose-colored, nodding flowers that look like they never completely open. They begin blooming in late April and continue well into summer, giving rise to showy mauve seed heads that resemble plumes of smoke. The ferny, light blue-green leaves are covered with soft hairs.
Landscape Use Prairie smoke is a good landscape plant, offering interest throughout the growing season. Use it in prairie gardens, perennial borders, and rock gardens. It eventually forms a dense groundcover. The seed heads attract goldfinches and can be dried for flower arrangements.
Site Dry, average, well-drained soil in full sun to light shade.
Culture Prairie smoke is a tough plant that withstands bitter cold, high heat, and drought. Rhizomes should be divided every third or fourth year to alleviate overcrowding.

Good Companions In prairie gardens, combine it with pasque flower, butterfly weed, Canada columbine, prairie phlox, and bird's-foot violet. In perennial borders or rock gardens, it combines nicely with spring bulbs and low-growing perennials such as *Phlox stolonifera* (creeping phlox), perennial geraniums, and sedums.

Other Species *G. rivale* (purple avens) is native to bogs, low woods, and wet prairies in the northeastern quarter of Minnesota. It grows up to 2 feet tall with purple, bell-like flowers that turn to plumes of seeds. It is not as showy as prairie smoke, but its interesting blooms are a nice addition to moist woodland gardens. Zone 2.

Geum triflorum

Helenium autumnale var. *autumnale*
Autumn sneezeweed
Zone 3

Helenium autumnale 'Butterpat'

Native Habitat Low woods, prairies, and marshes in all but the far northeastern section of Minnesota.

Height 3 to 5 feet

Description Sneezeweed has abundant golden-yellow, daisylike flowers 1 to 2 inches wide July through September. The bright green leaves are lance-shaped with toothed edges.

Landscape Use Autumn sneezeweed provides nice late-summer color in perennial borders or prairie gardens. It does best in a low, moist area such as bogs or near streams, but it can be grown in many landscape situations.

Site Prefers a dampish spot in gardens, perhaps slight depressions that can be given a good soaking during dry periods. Responds well to good, fertile soil.

Culture Although native to moist sites, sneezeweed adapts readily to most garden soils, especially if there is a low spot. Plants in moist, rich soil are quite robust, while those grown in drier soils are shorter and less vigorous. Taller plants may need some sort of support to keep from flopping over. Mulch garden plants and give them extra water during dry times, especially in midsummer, to encourage good flowering. Prune back plants in late May to keep them smaller and more compact. Plants will bloom better if they are divided every three or four years.

Good Companions Plant autumn sneezeweed near ponds with iris, ferns, New England aster, and ironweed. In borders, use it with phloxes, asters, coreopsis, and goldenrods.

Cultivars Several cultivars are available, but they are not all easy to locate. 'Butterpat' grows to only 4 feet with bright yellow flowers. 'Moerheim Beauty' has bronze-red blossoms. Zone 3.

Heliopsis helianthoides var. *scabra*

'Summer Sun'

Heliopsis helianthoides var. *scabra*
Oxeye
Zone 3

Native Habitat Woodland edges, open woods, and prairies throughout most of Minnesota.

Height 3 to 5 feet

Description Oxeye is a rather coarse plant with large, rough, medium-green foliage. It's redeemed by its abundance of sunny yellow, sunflower-like flowers that appear from June into September.

Landscape Use Oxeye is an excellent landscape plant, providing midsummer color in prairie gardens and perennial borders.

Site Moist or dry, average to rich soil in full sun or light shade.

Culture Oxeye adapts readily to garden culture. It is easy to grow from seed, often flowering the first summer if started indoors in winter. Plants require watering during dry periods to prevent wilting. Plants may get floppy; pinch them back in late May to reduce overall height. The named cultivars are less floppy. Oxeye will self-seed but seedlings are easily weeded out. It is occasionally attacked by aphids and powdery mildew.

Good Companions Plant oxeye with other summer bloomers such as blazing stars, butterfly weed, prairie phlox, wild bergamot, white wild indigo, garden lilies, daylilies, veronicas (*Veronica* species), asters, and grasses.

Cultivars Several cultivars are available, with 'Summer Sun' being the most popular and easiest to locate. It is more compact, growing to about 3 feet with large flowers. 'Prairie Sunset' is a newer introduction with bright yellow flowers with contrasting orange-red centers. Plants grow to 6 feet with attractive purplish stems and purple-veined leaves. Zone 4.

Hepatica americana
Round-lobed hepatica
Zone 3

Native Habitat Upland woodlands mainly in the northeastern half of Minnesota.

Height 4 to 6 inches

Description Round-lobed hepatica is one of the earliest flowers of spring, beginning its show in early April and continuing with scattered blooms into May. It has fuzzy, leafless stalks bearing pink, lavender, blue, or white flowers, each about 1 inch wide. Basal leaves are leathery and wine-colored on the underside with three distinct rounded lobes. They keep plants interesting after the flowers fade.

Landscape Use Plant hepaticas where you can enjoy their early bloom. A backdrop of rocks or a large tree trunk will help to set off the early flowers. Plants will multiply to carpet the ground in woodland gardens. Hepaticas are an attractive addition to shaded rock gardens.

Site Moist, humus-rich soil in light to full shade.

Culture Plants self-sow and clumps get bigger every year but never to the point of being invasive. Keep them away from larger, more-aggressive species that can overtake them. Round-lobed hepatica responds well to light fertilization in spring and an occasional application of limestone.

Good Companions Plant hepaticas with other woodland plants such as wild ginger, bloodroot, trilliums, rue anemone, Dutchman's breeches, and spring beauty. They bloom at the same time as early species crocus, snowdrops, *Scilla siberica* (Siberian squill), and *Puschkinia* (striped sqill) and are nice companions for these minor bulbs.

Cultivars and Other Species 'Louise' was selected from a garden in Owatonna, Minnesota. The double outer petals start out bright pink and the smaller center petals are slightly greenish. As flowers age, they turn pale blue. Zone 3.

Hepatica americana

H. acutiloba (sharp-lobed hepatica) is native to rich, deciduous woods in acidic soil in southeastern Minnesota. It grows 3 to 6 inches tall with dainty white or bluish flowers in early April. The almost-evergreen leaves have three sharp-pointed lobes. It likes a more acidic soil than *H. americana*, and so prefers a soil acidifier to an application of limestone, but otherwise is similar in culture and use. Zone 3.

Iris versicolor
Northern blue flag
Zone 2

Native Habitat Wet prairies, ponds, shallow marshes, and bogs throughout most of Minnesota.

Height 2 to 3 feet

Description This emergent aquatic is a rhizomatous perennial. From May to July, it produces several bluish violet flowers on a stout stem among a basal cluster of swordlike leaves.

Landscape Use Blue flag's flowers are attractive in early summer, and the erect, swordlike foliage provides textural contrast and a tropical feel throughout the growing season. Plant blue flag in a damp to wet spot, such as next to ponds or in bogs, where the leaves will lend a strong vertical accent. It can be grown in heavy soil in perennial borders where there is an ample water supply. Plants can be potted and sunk into wetland ponds.

Site Requires wet, rich soil in full sun to light shade. Grows best when the rhizome is just covered with water.

Culture If necessary, amend soil with organic matter to improve its water-holding capacity. Blue flag creeps slowly to form nice clumps, which are easily divided every third year for maximum bloom. To grow blue flag in water gardens, plant it in rich clay soil in a container. Cover the soil with 2 inches of pea gravel and submerge the pot in up to 8 inches of water.

Good Companions Grow blue flag with other moisture-lovers such as turtlehead, obedient plant, ferns, swamp milkweed, wild calla (*Calla palustris*), pickerelweed (*Pontederia cordata*), arrowhead, and marsh marigold.

Cultivars and Other Species 'Kermesina' has bright red-purple petals with white centers. 'Version' is a pink selection. Zone 2.

Iris versicolor

I. virginica var. *shrevei* (southern blue flag) is native to pond shores and marshes in southeastern Minnesota. It has slightly darker flowers on shorter stems and narrower leaves. The new growth has an attractive burgundy tinge that persists into summer. Cultural requirements and uses are the same. Zone 4.

Isopyrum biternatum
False rue anemone
Zone 3

Isopyrum biternatum

Native Habitat Rich deciduous woods and streamsides in the southeastern quarter of Minnesota.

Height 8 to 12 inches

Description False rue anemone has delicate, blue-green basal and stem leaves with small, round-lobed leaflets. Snow-white flowers appear in sparse clusters covering plants for a month in early spring. It is one of the earliest wildflowers to bloom, starting in early to mid April. Plants go dormant after blooming, with new leaves often appearing in fall.

Landscape Use This early spring bloomer makes a nice groundcover in shade or woodland gardens. It weaves its way among other flowers without overpowering them.

Site Humus-rich, evenly moist soil in light to full shade.

Culture Amend soil with organic matter before planting, if needed, to create the necessary growing conditions. Plants grow from creeping roots to spread, but they rarely become a nuisance. Plants disappear after flowering but new foliage may emerge in fall or late winter. This native flower may be difficult to locate in the nursery trade.

Good Companions Plant false rue anemone with other shade-loving plants such as wild ginger, blue phlox, bloodroot, trillium, primroses, epimediums, and pulmonarias.

Jeffersonia diphylla
Twinleaf
Zone 4

Jeffersonia diphylla

Native Habitat Deciduous forests, usually in sheltered ravines preferring north- and east-facing slopes in far southeastern Minnesota.

Height 6 to 15 inches

Description In early spring this perennial has a solitary, eight-petaled, white flower up to 2 inches wide on a leafless stalk. The flowers last only a day or two, but the plant redeems itself with its interesting foliage. The long-stemmed basal leaves are deeply divided into two lobes, resembling a perched butterfly, and they have an attractive reddish tinge. Flowers are followed by interesting pipelike seed capsules in early summer.

Landscape Use Twinleaf can be used as groundcover in shade or woodland gardens where the attractive leaves overlap and form layers.

Site Prefers an evenly moist, rich, neutral to alkaline soil in spring sun but will tolerate fairly dry soils.

Culture Add limestone to the soil if the pH is too acidic. This plant is on Minnesota's special concern list. Growing nursery-propagated specimens will help the species, but do not use wild-dug plants.

Good Companions Grow twinleaf with other deciduous woodland plants with interesting foliage such as bloodroot, ferns, meadow rue, trout lilies, and wild ginger. In its native habitat, it is often associated with other rare species such as golden seal (*Hydrastis canadensis*) and squirrel corn (*Dicentra canadensis*).

Liatris pycnostachya
Great blazing star
Zone 3

Native Habitat Moist black-soil prairies and moist savannas mainly in the southern two-thirds of Minnesota.

Height 3 to 5 feet

Description Great blazing star has showy red-violet to mauve terminal flower spikes that crowd the upper portion of the stiff, leafy stems. The alternate, grasslike leaves increase in size from top to bottom. It blooms in midsummer, with the bloom starting at the top of the flower spike and working its way down.

Landscape Use Blazing stars are wonderful plants for perennial borders and prairies gardens. Bees, hummingbirds, and monarchs and other butterflies gather on the flowers all summer, and birds eat the seeds. Great blazing star is a good cut flower and holds its color well when dried.

Site Moist, fertile, well-drained soil in full sun.

Culture Great blazing star is somewhat drought resistant once established. The tall stems usually need support, which can come from staking or nearby grasses. Plants reseed but never become weedy and seldom need dividing. Cut back plants in spring rather than fall so birds can feast on the seed heads. Pocket gophers, mice, and voles like to eat the corms, and rabbits nibble on young foliage.

Good Companions Plant blazing stars with other summer prairie plants such as Culver's root, rattlesnake master, wood lily, mountain mint, coneflowers, wild bergamot, goldenrods, and milkweeds.

Cultivars and Other Species 'Alba' [var. *alba*] has creamy white flowers.

L. aspera (rough blazing star) is native to dry prairies and savannas across the southwestern half of Minnesota. It has clusters of 1-inch, pale purple or pink, buttonlike flowers on short stalks on top of 3- to 5-foot stems in mid to late summer. The leaves are gray-green and grasslike. Grow it in sandy or loamy, moist but well-drained soil in full sun in prairie gardens or mixed borders. It will need staking or the support of nearby plants. Birds and butterflies enjoy the flowers. Zone 3.

L. cylindracea (cylindric blazing star) is native to dry prairies, savannas, and open woods mainly in southeastern Minnesota but also farther west and north. It is shorter, growing 8 to 24 inches, with narrow clusters of pale purple flowers in open spikes in late summer. Culture and use are the same as great blazing star. Zone 3.

L. ligulistylis (northern plains blazing star) is native to wet black-soil prairies and borders of marshes in all but far northeastern Minnesota. It looks similar to cylindric blazing star but with dark violet flowers that are broader and more open. It grows 3 to 5 feet tall and requires more soil moisture. Use it in wet prairie gardens and alongside streams and ponds. It is a favorite of monarch butterflies. Zone 3.

L. punctata (dotted blazing star) is native to dry, sandy prairies mainly in western and central Minnesota. It grows 8 to 14 inches tall with pink-purple wands of dense flowers and silvery gray leaves. It is tolerant of hot, dry conditions. Zone 3.

Liatris pycnostachya

Liatris aspera

Lilium michiganense
Michigan lily
Zone 4

Native Habitat Moist to wet soils mainly in the eastern half of Minnesota.

Height 3 to 4 feet

Description Michigan lily has nodding, deep orange flowers with strongly recurved petals flecked with brown in July to August. The 2- to 3-inch flowers are held nicely above the stems and the whorled leaves have smooth margins.

Landscape Use Michigan lily can be used in low areas in perennial borders or along ponds and streams. It is a good bog plant. Hummingbirds enjoy the flowers.

Site Wet, rich soil in full sun to light shade.

Culture Michigan lily likes a deep, rich soil, so add lots of organic matter before planting. It is a stoloniferous plant, with bulbs growing at the ends of the rhizomes. Bulbs should be planted in fall. Rodents often eat the bulbs; deter them by surrounding planted bulbs with 1 to 2 inches of gravel.

Good Companions Plant Michigan lily with other moisture-loving natives such as turtlehead, northern plains blazing star, ferns, and Joe-pye weeds.

Other Species *L. philadelphicum* var. *andinum* (wood lily) is native to well-drained loams in prairies and open woodlands throughout most of Minnesota. It has upright, 2-inch, cup-shaped, orange-red flowers spotted with pur-

Lilium michiganense

plish brown, and erect, 1- to 3-foot stems along with whorled, widely space, lance-shaped leaves. It is one of the few lily species with upward-pointing flowers, and blooms June to August. Grow it in dry, average, acidic, well-drained soil in full to part sun. Plants are difficult to establish and tend to be short-lived. Zone 3.

Lobelia cardinalis var. cardinalis
Cardinal flower
Zone 3

Native Habitat Moist to wet soils in open wet woods, swamps, and sandbars in streams along the Mississippi and St. Croix Rivers in eastern Minnesota.

Height 24 to 48 inches

Description Cardinal flower gets its name from the brilliant rich red flowers that grow in an elongated cluster atop the stems, July to September. Leaves are alternate, dark green, and canoe shaped, 2 to 6 inches long.

Landscape Use Cardinal flower is one of the few native plants with true red flowers, and they are a welcome addition to late-summer landscapes. Plant it in groups of five to seven in a moist area of perennial borders or woodland gardens. A dark background will set off the flowers nicely. It thrives at the edges of water gardens and in bogs and will naturalize when conditions are right. The showy tubular flowers attract hummingbirds.

Site Moist to wet, average soil in partial sun. It will tolerate full sun as long as the soil is always at least slightly damp.

Culture Cardinal flower adapts well to gardens, despite its native tendency to

grow in moist areas at the edges of water. Amend soil with lots of organic matter and peat moss before planting. It transplants easily, but is short-lived, so add seedlings every couple of years. Plants may not survive the winter if there is insufficient soil moisture; winter mulch is helpful.

Good Companions A backdrop of ferns will help set off the red flowers. Cardinal flower grows well with blue flags, autumn sneezeweed, astilbes, turtleheads, and ligularias.

Other Species and Cultivars

L. siphilitica (great lobelia) grows along bottomlands and riverbanks across the southwestern half of Minnesota. It has pretty, bright blue flowers in leaf axils on the upper portion of leafy, 1- to 3-foot stems. It blooms for a long time, July to September. Give it a moist, rich soil in sun to light shade. It brings a welcome shade of blue to late-summer gardens. In the right conditions, it reseeds readily, but the seedlings are easy to pull. 'Alba' is a white-flowered form. Zone 3.

L. spicata (pale-spiked lobelia) grows in moist prairies, clearings, and

Lobelia cardinalis

Lobelia siphilitica

marsh edges across southwestern Minnesota. It is a biennial with pale blue flowers in elongated clusters on top of 24- to 40-inch leafless stalks in summer. It is more tolerant of dry, poor soil and grows in full sun to light shade. Plants will self-sow once established. Zone 3.

Lupinus perennis
Wild lupine
Zone 3

Native Habitat Dry, sandy soils in a band from east-central to southeastern Minnesota.

Height 15 to 30 inches

Description Wild lupine has pealike, blue to purple flowers that grow in a 12-inch terminal cluster on erect, unbranched stems. Alternate leaves are palmately compounded, each with seven to eleven leaflets measuring 1 to 2 inches long. It blooms mid-May into June.

Landscape Use A mass planting of wild lupine is a stunning sight. Plant groups in full sun in prairie gardens and woodland borders or clearings. It can be used in perennial borders, but plants often go into dormancy after flowering, so surround them with late-blooming plants. It is the only food plant for the caterpillar of the endangered karner blue butterfly.

Site Prefers dry, slightly acidic, well-drained soil in full sun or light shade. Thrives in poor soils and summer drought.

Culture Wild lupine can be a bit difficult to establish in landscape settings. Do not attempt to transplant mature plants. Start with young seedlings, and include some of the soil from their mother plant with them to ensure the presence of specific nitrogen-fixing bacteria associated with the roots. Plants can also be started from seeds, which should be inoculated with the appropriate bacteria before planting. Once established, wild lupine enhances soil fertility by fixing nitrogen from the atmosphere. Space plants 8 to 12 inches apart in groups of five to seven. Plants will be short-lived on sites not suited to them.

Good Companions Although wild lupine looks best in large masses in naturalized settings, you can plant individual specimens with phloxes, spiderworts, blazing stars, and grasses in perennial borders.

Lupinus perennis

Maianthemum canadense
Canada mayflower
Zone 2

Native Habitat Moist woodlands, lakeshores, and clearings in all but southwestern Minnesota.

Height 2 to 4 inches

Description Canada mayflower has zigzag stems topped by 1- to 2-inch racemes of minute white flowers resembling lily of valley. It blooms mid May into June. Plants have 1- to 3-inch, alternate leaves with heart-shaped bases and smooth margins that clasp the stem. Flowers turn to reddish berries that are food for birds.

Landscape Use Canada mayflower makes a nice groundcover of glossy green in shade gardens. Under favorable conditions, it spreads to cover large areas, weaving itself among nearby plants.

Site Requires moist, acidic, humus-rich soil in part to deep shade.

Maianthemum canadense

Culture Acidify soil before planting, if necessary. Mulch with pine needles or shredded leaves. Canada mayflower spreads readily from rhizomes when conditions are right. It doesn't like to be crowded by other plants.

Good Companions Plant Canada mayflower in shade gardens with other acid-loving plants such as partridge-berry, starflower, bunchberry, twinflower, and round-lobed hepatica.

Mertensia virginica
Virginia bluebells
Zone 3

Mertensia virginica

Native Habitat Moist deciduous woods, floodplains, streamsides, and clearings in the southeastern corner of Minnesota.

Height 12 to 24 inches

Description Clusters of pinkish buds open to showy, drooping clusters of sky-blue, trumpet-shaped flowers. The flowers begin opening in late April and last a long time, but plants go dormant soon after blooming. The lettucelike leaves are oval-shaped, thick-veined, and deep green with smooth margins. They are a showy deep purple color in early spring when they first emerge.

Landscape Use Virginia bluebells are easily grown in woodland or shade gardens, where the flowers provide a soothing sea of blue for many weeks. They can also be used as a weaver in semi-shady perennial gardens, filling in spaces before other perennials are up and blooming. Gently pull mulch away from emerging plants so you can enjoy the striking deep purple color of the new foliage.

Site Prefers moist, rich sites in part sun or shade.

Culture Plants die back after flowering in June, so interplant them with ferns or wild ginger. Be careful not to dig into dormant clumps. They will self-seed but not to the point of becoming a nuisance. Plants in perennial borders should be mulched and fertilized.

Good Companions Plant this early bloomer with yellow- or orange-flowered spring bulbs in perennial borders. In woodland gardens, combine it with celandine poppy (*Stylophorum diphyllum*), large-flowered bellwort, shooting star, columbines, and trillium. Use wild ginger and ferns to fill in after Virginia bluebells go dormant.

Mitella diphylla
Two-leaved miterwort, bishop's cap
Zone 3

Mitella diphylla

Native Habitat North-facing limestone or sandstone bluffs and rocky slopes mainly in southeastern Minnesota but also farther north.

Height 6 to 12 inches

Description Miterwort is a neat and compact rhizomatous perennial with evergreen, toothed, triangular leaves that stay low to the ground. Flowers are small and dainty in terminal racemes on top of 6- to 12-inch-tall stalks. They are creamy white to green or purplish and bloom early May into June.

Landscape Use Miterwort spreads slowly but eventually makes a nice groundcover under shrubs such as viburnums, leatherwood, and dogwoods, or in woodland gardens. A generous planting of miterwort in flower has the same effect as baby's breath in a flower arrangement.

Site Prefers a humus-rich, moist but well-drained, neutral or slightly acidic soil in light to full shade.

Culture Amend soil with organic matter before planting. Space plants 8 to 10 inches apart. Miterwort is surprisingly tolerant of drought, and will actually grow on top of limestone with only a few inches of rich soil.

Good Companions Grow miterwort in masses with columbines, woodland phlox, celandine poppy (*Stylophorum diphyllum*), bleeding heart, Jacob's ladder, and wild geranium.

Monarda fistulosa
Wild bergamot
Zone 3

Native Habitat Prairies, clearings, and savannas throughout most of Minnesota.

Height 2 to 4 feet

Description Wild bergamot has soft lavender to pale pink, 3-inch, tubular flowers in dense, round, terminal clusters. It blooms June to August. The foliage is rather coarse and the stems are square, a mint-family characteristic.

Landscape Use Wild bergamot is a beautiful summer-blooming perennial that adapts well to landscape use. It can be used in prairie plantings and in the middle to back of perennial borders. The flowers attract bees, butterflies, and hummingbirds.

Site Prefers average to rich, well-drained soil in full sun to light shade.

Culture Wild bergamot tolerates varying soil fertility as long as it is well drained. If the soil is too rich, stems can become weak. Plantings have a tendency to die out in the middle. Dividing plants every three to four years helps keep them vigorous and reduces their spread. Powdery mildew may be a problem in wet, humid conditions, but it is rarely serious. Avoid overhead watering.

Good Companions The soft color and uniquely shaped flower of wild bergamot is a nice complement to the brighter, deeper colored flowers of other summer-blooming prairie plants such as black-eyed Susan, blazing stars, and butterfly weed. In perennial gardens, plant it with garden phlox (*Phlox paniculata*), yarrows (*Achillea* species), perennial geraniums (*Geranium* species), and ornamental grasses.

Monarda fistulosa

Nymphaea odorata
American white water lily
Zone 2

Native Habitat Ponds, sloughs, marshes, and swamps in east-central and northeastern Minnesota.

Height Floats on water surface

Description This rhizomatous aquatic has leaves and flowers that float on the water surface in areas of lakes and ponds up to 5 feet deep. Flowers are many-petaled, pure white or rarely pinkish, 3 to 5 inches wide with a multitude of yellow stamens. They open in the morning and close in the afternoon. Most flowers are fragrant. The round, flat, shiny green leaves are 4 to 12 inches across and usually purple on the backside.

Landscape Use American white water lily can be grown in large water gardens and ponds, where plants will spread to create a solid mat. Container-grown plants can be grown in small ponds and tub gardens.

Site Requires a submerged, rich soil in full sun.

Culture American white water lily grows in deep ponds with the rhizomes barely covered with soil. In most landscape situations, they should be grown in pots to keep them from spreading too aggressively. Partially fill dark pots with clay topsoil low in humus and plant rhizomes 3 to 4 inches deep. Cover the soil surface with 2 inches of fine gravel to hold in the topsoil. Keep the newly potted container within 6 to 8 inches of the water surface until several new leaves appear, then move it to water up to 4 feet deep. If planted in small, shallow garden ponds that may freeze solid, bring potted plants indoors in late fall, wrap each one in wet newspaper, seal the package in a plastic bag, and store in a cold basement where the temperatures remain a few degrees above freezing.

Good Companions Grow American white water lily with pickerelweed, (*Pontederia cordata*) and arrowhead (*Sagittaria latifolia*), two other native emergent aquatics.

Nymphaea odorata

Opuntia humifusa
Eastern prickly pear
Zone 4

Native Habitat Sandy prairies in southern Minnesota.

Height 8 to 12 inches

Description Eastern prickly pear has large yellow flowers 2 to 3 inches wide with or without reddish centers that bloom June to July. The stems have been modified into flattened, fleshy pads covered with clusters of short bristles. The barrel-shaped fruits turn reddish purple as they ripen.

Landscape Use Their alien look and cultural conditions limit how prickly pears can be used in landscapes, but they are great in rock gardens, stone walls, and xeriscapes.

Site Well-drained, sandy soil in full sun.

Culture Prickly pears are difficult to weed and cultivate around. The tiny spines can be irritating; wear heavy gloves when handling or working around plants. They survive winter temperatures by withdrawing most of the moisture from their pads, giving them a shriveled, unhealthy look in winter and early spring. They quickly plump up in spring, however. The clump-forming plants are easily cut back if they become too sprawling. It is on Minnesota's special concern list, so only plant nursery-propagated specimens.

Good Companions Grow prickly pears with other plants that like the same hot, dry conditions such as beard-tongues, wild onions, harebell, purple prairie clover, rattlesnake master, and flowering spurge.

Other Species *O. macrorhiza* (plains prickly pear) is native on thin soils over bedrock in southwestern Minnesota. It is a coarse, sprawling plant with large, flattened joints. Culture and use are the same. It is on Minnesota's special concern list, so only purchase nursery-propagated plants. Zone 4.

Opuntia humifusa

Penstemon grandiflorus
Large-flowered beardtongue
Zone 3

Native Habitat Prairies, woodland edges, and embankments mainly in central and western Minnesota.

Height 1 to 4 feet

Description Large-flowered beard-tongue has tubular, deep lavender flowers with flat faces in late May or early June. The large, oval leaves are deep green and untoothed.

Landscape Use Beardtongues can be grown in prairie gardens or perennial borders. Since plants have a short bloom time and the foliage isn't all that attractive, plant later-blooming plants nearby to distract from nonblooming plants. Their tolerance of dry soil makes them well suited to rock gardens. The flowers are attractive to hummingbirds.

Site Average to rich, sandy or loamy, well-drained soil in full sun.

Culture This short-lived species will self-sow to replenish itself. A fertile soil encourages heavier bloom. Plants will not grow well on heavy clay soils.

Good Companions Plant beardtongue with other late-spring bloomers such as wild lupine, geraniums, yarrows, ornamental onions, prickly pears, garden phlox, and Siberian iris.

Other Species *P. gracilis* (slender beardtongue) is native to open woods, savannas, and prairies throughout most of Minnesota. It is a more delicate plant, with smaller, rose-pink to purple flowers on 2-foot stems. The narrow, toothed leaves are covered with soft hairs. Culture and use are the same. Zone 3.

Penstemon gracilis

Phlox divaricata
Blue phlox, woodland phlox
Zone 3

Native Habitat Rich mesic woods, moist deciduous woods, clearings, and flood plains in southern Minnesota.

Height 10 to 20 inches

Description Blue phlox has clusters of fragrant, pretty, pale blue to dark purple-violet flowers late April into June. The glossy, semi-evergreen foliage spreads at a moderate rate by creeping rhizomes that form loose mats.

Landscape Use This is one of the best groundcovers for woodland gardens and is ideal for underplanting deciduous trees and large shrubs where it will receive early spring sun and summer shade. Use blue phlox to hide dying spring bulb foliage. It forms natural drifts and is good for covering hillsides.

Site It thrives in partial to full shade in cool, well-drained soil that is rich in organic matter.

Culture Blue phlox has a shallow root system and benefits from a summer mulch to conserve soil moisture. Flowering will diminish and foliage will brown if conditions are too sunny or too dry. Cut back flower stalks after flowering to keep plants neat. Phloxes are favorite foods of deer and rabbits. To revive eaten plants, fertilize with a mixture of 1 tablespoon fish emulsion in 1 gallon of water. Plants seldom need dividing.

Good Companions Plant blue phlox with spring bulbs and wildflowers such as trillium, celandine poppy (*Stylophorum diphyllum*), columbines, Virginia bluebells, spring beauties, and wild ginger. In woodland gardens, plant it with foamflower (*Tiarella cordifolia*), bleeding hearts, and ferns.

Cultivars and Other Species Var. *laphamii* has deeper blue flowers. 'Fuller's White' is a good white-flowered cultivar. 'Dirgo Ice' has pale blue to white flowers. 'Clouds of Perfume' has fragrant ice-blue flowers. 'Chattahoochee' is a hybrid of *P. divaricata* var. *laphamii* and *P. pilosa* with lavender-blue flowers with dark purple centers.

P. maculata (wild sweet William) is native to moist woods and wet prairies in southeastern Minnesota. It grows 2 to 3 feet tall with large, conical clusters of pink or white flowers. The lance-shaped green leaves look nice all season. It prefers drier, sunnier conditions than its woodland cousin. Plant it in average to rich, moist but well-drained soil in full sun or light shade in perennial borders or prairie gardens with black-eyed Susan, prairie smoke, nodding wild onion, golden alexanders, wild bergamot, and mountain mint. Divide the multistemmed clumps every three to four years to keep plants vigorous. 'Alpha' has rose-pink flowers with darker eyes. 'Rosalinde' has deep pink flowers and grows 3 to 4 feet tall. Zone 3.

P. pilosa var. *fulgida* (prairie phlox) is native to prairies, dry open woodlands, and savannas in the southern two-thirds of Minnesota. It grows 10 to 30 inches tall with clusters of lavender, pink, or sometimes whitish, fragrant flowers atop downy stems and leaves. It blooms late spring and early summer. Grow it in sandy, well-drained, slightly to moderately acidic soil in full sun or light shade in prairie gardens or perennial borders. Zone 3.

Phlox divaricata

Phlox maculata

Phlox pilosa

Physostegia virginiana var. virginiana
Obedient plant
Zone 3

Physostegia virginiana

Physostegia virginiana 'Vivid'

Native Habitat Moist prairies, low wet woods, bottomlands, and borders of marshes throughout most of Minnesota.

Height 2 to 3 feet

Description Obedient plant has tall vertical stems lined with narrow, jaggedly toothed leaves. The showy tubular flowers bloom midsummer to early fall. They appear successively up the stalk in somewhat elongated clusters of pinkish, two-lipped, inch-long, snapdragon-like flowers arranged in two rows.

Landscape Use Obedient plant's aggressive nature limits its landscape use. Use it in wet prairie gardens and naturalized plantings. The flowers have an old-fashioned look to them and are good for Victorian gardens and cut bouquets. Several good cultivars are available that are less aggressive and better for garden use.

Site Prefers a moist, rich, well-drained soil in full sun but tolerates drier conditions and partial sun.

Culture Plants spread by underground rhizomes and can become invasive. In perennial borders, plants can be grown in buried 3-gallon nursery containers to keep them in bounds. Garden plants may need staking, especially if grown in fertile soils.

Good Companions Obedient plant's pink to white flowers are a welcome addition to late-summer borders. Plant it with New England aster, bottle gentian, blazing stars, Culver's root, ironweed, Joe-pye weeds, goldenrods, and grasses.

Cultivars 'Miss Manners' is much less aggressive than the species, well behaved enough to be used in perennial borders. It has bright white flowers late summer through fall. 'Pink Bouquet' has bright pink flowers on 3- to 4-foot stems. 'Variegata' has pale pink flowers and leaves edged in creamy white. 'Vivid' has vibrant rose-pink flowers on 2- to 2½-foot stems.

Podophyllum peltatum
May apple
Zone 3

Native Habitat Moist open woods and clearings in the southeastern corner of Minnesota.

Height 12 to 18 inches

Description May apple is a rhizomatous, spreading perennial with nodding, creamy white, 2-inch flowers mid to late May. The single flower is hard to see beneath the large, umbrella-like, deeply lobed, paired leaves that spread up to a foot in diameter.

Landscape Use May apples are good for covering large shady areas of bare ground; their large, tropical-looking leaves emerge in mid-April and they remain fresh-looking all season in cooler temperatures. Their aggressive nature can overtake more delicate species, so they are usually planted in large masses. If you want to see the spring flowers, plant it on slopes so you can look up under the leaves.

Site Grows in evenly moist to damp, humus-rich soil in light to full shade but tolerates drier sites once it is established.

Culture Once established, May apple spreads rapidly by long rhizomes and may crowd out nearby plants. It can be kept under control with annual removal of some rhizomes, which are easily dug. If you want to grow it with less-aggressive plants, plant it in a large sunken nursery pot. The green, applelike fruits are poisonous until they are fully ripe, as are all other parts of the plant.

Podophyllum peltatum

Good Companions Few plants can compete with May apple, and it is usually grown as a large groundcover. In large shaded areas, it can be grown with other large plants such as Solomon's seal, baneberries, bugbane (*Cimicifuga racemosa*), and large ferns. It can also be grown with early ephemerals that go dormant before it shades them.

Polemonium reptans
Spreading Jacob's ladder
Zone 3

Native Habitat Neutral soils of moist deciduous woods and floodplains mainly in southeastern Minnesota.
Height 8 to 16 inches
Description Spreading Jacob's ladder is a low-branching, clumping plant with attractive, dark green, pinnately divided leaves. Plants are covered with small sky-blue flowers in terminal clusters held slightly above leaves starting in late April to June.
Landscape Use This long-blooming plant works well in woodland or shade gardens and under shrubs or flowering trees.
Site Prefers a fertile, damp soil with a pH of 6 to 7 in part shade, but can be grown in almost full sun.

Culture Cut bloom stalks to the ground after flowering; foliage will remain attractive all season. Plants do not creep and they seldom need division. The brittle flower stems are easily broken so any staking should be done before plants flower.
Good Companions Combine Jacob's ladder with Virginia bluebells to provide blue color after the latter goes dormant. It looks nice under flowering trees such as serviceberries or *Prunus nigra* 'Princess Kay'. In shade gardens, combine it with bloodroot, wild geranium, foamflower (*Tiarella cordifolia*), wild ginger, hostas, and ferns.
Cultivars 'Blue Pearl' stays at 10 inches tall and has abundant, rich blue flowers.

Polemonium reptans

Polygonatum biflorum
Giant Solomon's seal
Zone 3

Polygonatum biflorum

Native Habitat Rich, moist deciduous or mixed coniferous woods and floodplains in all but northeastern Minnesota.
Height 1 to 3 feet
Description Giant Solomon's seal has greenish yellow flowers that dangle from leaf axils in May and June, but the deep purple fruits that appear later in summer are much more ornamental. They hang from the arching stems and offer great contrast to the foliage, which turns a striking gold color in fall. The alternate, oval, 2- to 6-inch, parallel-veined leaves clasp the stem.
Landscape Use The graceful, arching stems of giant Solomon's seal bring a strong architectural element to gardens. Use it as groundcover at the base of large trees, where it tolerates the dry shade. Tuck clumps here and there for accent, setting off the bright green leaves with a tree trunk or fallen log. Birds eat the berries.
Site Prefers moist, high-humus soil in partial sun to deep shade but tolerates drier sites.
Culture Giant Solomon's seal grows slowly from creeping rhizomes to form

natural patches, but seldom becomes invasive. For faster growth, mulch with pine needles or oak leaves to increase soil acidity and moisture. Plants grown in sun may need watering during dry periods.
Good Companions Plant giant Solomon's seal with other tall foliage plants for an interesting textural display. Good choices are ostrich fern, American spikenard, May apple, and false Solomon's seal.
Cultivars and Other Species Var. *commutatum* (great Solomon's seal) is

a rare naturally occurring tetraploid form (having more than the regular number of chromosomes). It can grow up to 6 feet tall or more and is truly spectacular, but is difficult to find. Zone 3.

P. pubescens (hairy Solomon's seal) is native to rich woodlands mainly in the northeastern half of Minnesota. It is a little smaller, growing to only about 28 inches, and has hairy veins on the lower leaf surfaces. Culture and use are the same as giant Solomon's seal. Zone 3.

Pontederia cordata
Pickerelweed
Zone 4

Pontederia cordata

Native Habitat Marshes, streams, shallow lakes, and ponds mainly along the eastern half of Minnesota.

Height 1 to 2 feet above water surface

Description This emergent aquatic grows in shallow water (3 feet or less) or saturated soil near water features, often forming large colonies. It has attractive clublike clusters of violet-blue flowers (or rarely, white) June to August. Flowers emerge from a sheathing basal spathe with heart-shaped, glossy leaves 2 to 10 inches long.

Landscape Use Pickerelweed is an ornamental plant forming spreading patches by means of a thick, creeping rhizome. It can be grown in bog gardens or in shallow, large ponds. In smaller ponds, grow it as a container plant. It makes a nice cut flower. Ducks and geese eat the seeds, and muskrats may eat the leaves. This plant is the sole nectar source for one bee species (*Dufourea novae-angliae*).

Site Requires a fertile, heavy loam soil at water edges or a submerged soil in full sun but tolerates low fertility and partial shade.

Culture Once established, it is important to maintain year-round water depths greater than saturation but shallower than the leaves. Pickerelweed responds well to organic fertilizers. Plants slowly spread to cover the sediment with a tough vegetative mat, but they are easily weeded out if needed.

Good Companions Grow pickerelweed with American white water lily and arrowhead (*Sagittaria latifolia*), two other native emergent aquatics.

Pulsatilla [Anemone] patens
Pasque flower
Zone 2

Pulsatilla patens

Native Habitat Dry-soil and rocky prairies and savannas across the southwestern half of Minnesota.

Height 6 to 12 inches

Description Pasque flower has solitary lavender, blue, or white flowers rising above clusters of basal leaves starting in late April and lasting into May. The entire plant is covered with silky hairs that give it a silvery sheen. It blooms early enough that flowers often appear in matted remains of the previous year's foliage. Showy, long-tailed seed heads follow the flowers, extending the ornamental value.

Landscape Use Pasque flower is the earliest of the prairie wildflowers, making it a must for prairie gardens. Select a spot where it won't be overwhelmed by taller plants. It also does well in perennial borders or rock gardens.

Site Prefers a dry, well-drained soil in full sun to very light shade.

Culture Pasque flower is much tougher than it looks. The early flowers can handle late-spring snows and cold snaps. It needs moisture while flowering, but otherwise is quite drought tolerant. If soil is heavy, add sand and organic matter before planting. The nomenclature of this plant is confusing, and you will find it also listed as *Anemone patens*, *Pulsatilla nuttalliana*, or *P. hirsutissima*.

Good Companions Combine the demure pasque flower with other small spring bloomers such as minor bulbs, prairie smoke, pussytoes, bird's-foot violets, and columbines.

Pycnanthemum virginianum
Virginia mountain mint
Zone 3

Native Habitat Wet to mesic open woods and prairies mainly across the southwestern half of Minnesota.

Height 2 to 3 feet

Description Virginia mountain mint is a stiff, erect, clump-forming plant with leaves that smells like mint or oregano. The whitish to lavender flowers have purple spots and are borne in 1-inch terminal clusters in summer. The attractive 2-inch leaves have a bright yellow fall color and the gray seed heads cling to plants through winter.

Landscape Use Virginia mountain mint is a little coarse for perennial borders, but it can be used in informal wild gardens and prairies. It is often grown for its aromatic foliage rather than its flowers. It offers winter interest and attracts butterflies and bees. It makes a good cut or dried flower.

Site Prefers moist, humus-rich soil in full sun.

Culture This stoloniferous plant can become invasive. If you want to grow it in butterfly or herb gardens, plant it in large sunken nursery containers. Plants need division every few years to keep them from overwhelming nearby plants.

Good Companions Grow Virginia mountain mint with other summer bloomers such as purple prairie clover, blazing stars, black-eyed Susans, butterfly weed, prairie phlox, coneflowers, and beardtongues.

Pycnanthemum virginianum

Ratibida pinnata
Gray-headed coneflower
Zone 3

Native Habitat Dry prairies and savannas mainly in the southern third of Minnesota.

Height 3 to 5 feet

Description Gray-headed coneflower is a stiff, erect plant with rough-feeling, coarse leaves. It is redeemed by its showy 2½-inch flowers that appear throughout summer. They have drooping, soft-yellow rays and elevated, globose, conelike green centers that change to dark purple or brown.

Landscape Use The soft yellow color of gray-headed coneflower is a welcome and relaxing contrast to other hot-colored summer flowers. It adapts well to perennial borders and weaves nicely through prairie gardens. The flowers bloom a long time, attracting butterflies and bees, and are good for cutting. Once established, it is drought tolerant and can be used in xeriscapes.

Site Requires average to rich, well-drained soil in full sun or light shade. Plants cannot tolerate heavy, wet soils.

Culture Gray-headed coneflower is easy to grow. It transplants readily and seldom needs division. Plants reseed in the garden and prairie. Seedlings can be replanted to refresh the supply of plants. Garden plants may need staking in fertile soils.

Good Companions Plant coneflowers with other summer-blooming plants such as blazing stars, blanket flower, butterfly weed, purple prairie clover, wild bergamot, blue giant hyssop, and prairie grasses.

Other Species *R. columnifera* (Mexican hat, prairie coneflower) is native to dry prairies and savannas mainly in western Minnesota but also farther east. It is smaller, growing 1 to 3 feet tall. The taller center cones (1- to 1½-inch) and 1-inch drooping rays give the flowers a look of sombreros. Use and culture are the same. Zone 3.

Ratibida pinnata

Ratibida columnifera

Rudbeckia hirta var. *pulcherrima*
Black-eyed Susan
Zone 2

Rudbeckia hirta
var. pulcherrima

Rudbeckia hirta
'Indian Summer'

Rudbeckia triloba

Native Habitat Prairies and open woodlands throughout Minnesota.

Height 1 to 3 feet

Description Common black-eyed Susan may well be the most recognizable native flower. It is a biennial or short-lived perennial with large, flat flowers with golden-yellow rays surrounding a conical cluster of rich brown disc florets. The thin, lance-shaped leaves are covered with prickly hairs.

Landscape Use Black-eyed Susans bring long-lasting color to prairie gardens and mixed borders. Plants bloom the second year after seeding a prairie, offering color before most other prairie plants. The flowers attract bees and butterflies and provide winter interest if left standing. It is one of the few native flowers that adapts well to container culture, especially the cultivars.

Site Prefers moist, average, well-drained soil in full sun but tolerates light shade and a wide range of soil conditions.

Culture Sow seeds two years in a row for yearly blooming. Once a planting is established, plants will self-seed to keep the color coming year after year. Rich soils tend to produce weak-stemmed plants. Plants are easy to propagate from seeds or by transplanting seedlings.

Good Companions Good native companions include blazing stars, obedient plant, asters, butterfly weed, wild bergamot, little bluestem, and prairie drop seed. The list of suitable nonnative companions includes Russian sage (*Perovskia*), garden phlox (*Phlox paniculata*), sedums, and ornamental grasses such as *Calamagrostis* and *Pennisetum*.

Cultivars and Other Species 'Indian Summer' is a popular tetraploid form with larger flowers. 'Prairie Sun' has 5-inch flowers with golden petals tipped in primrose yellow surrounding a light green cone. 'Cherokee Sunset' is a mix of fully double flowers in shades of yellow, orange, bronze, and russet. All three cultivars are short-lived and are usually grown as annuals. They make good container plants. Showy tetraploid hybrids called gloriosa daisies are derived from *R. hirta*. They are annuals or short-lived perennials with 5- to 6-inch, yellow, orange, red, or multicolored flowers on 2- to 3-foot plants. They bloom all summer and are good for cutting.

R. laciniata var. *laciniata* (tall coneflower) is native to moist open areas throughout most of Minnesota. It grows 3 to 6 feet tall and has leaves with cut, irregular lobes. The abundant clusters of yellow flowers have light green central disks. It requires moister soil conditions and is great for soggy stream banks. Use it in informal gardens and wet prairies. Plants can take light shade, but with fewer flowers. 'Gold Drop' is a 2- to 3-foot-tall selection with double flowers that is good for perennial borders and containers. Zone 3.

R. triloba var. *triloba* (three-leaved coneflower) is native to moist prairies and open woodlands. It grows 2 to 5 feet tall, blooming July into October. It has brilliant yellow, smaller-sized flowers with jet-black centers and three-lobed leaves. It is a short-lived perennial that blooms its second year from seeding and self-sows readily in open soil. Use it in prairie gardens, perennial borders, and containers. It is on Minnesota's special concern list, so only purchase nursery-propagated plants. Zone 3.

Ruellia humilis
Wild petunia
Zone 4

Native Habitat A dry prairie in a single population in southern Washington County in east-central Minnesota.

Height 1 to 2 feet

Description Wild petunia has hairy stems and 3-inch stalkless leaves. The tubular, violet-blue flowers resemble annual petunias and bloom June to August.

Landscape Use Wild petunia is a great plant for dry, rocky, shallow soils such as in rock gardens, but it can also be grown in well-drained garden loam. Keep it at the front of perennial borders so larger plants don't overwhelm it. The lovely flowers bring the rare blue color to prairie gardens.

Site Needs well-drained soil in full sun.

Culture Wild petunia is easily grown in gardens if the soil is well drained. Add sand and organic matter to heavier soils. Plants grow from fibrous-rooted crowns to form clumps but never become invasive. Unfortunately, it is a favorite of deer.

Good Companions Plant wild petunia with other small plants that won't hide its showy flowers. Good choices include prairie smoke, wild onions, pussytoes, little bluestem, and butterfly weed.

Ruellia humilis

Sagittaria latifolia
Broad-leaved arrowhead
Zone 3

Native Habitat Along stream margins and in still ponds throughout Minnesota.

Height 12 to 36 inches

Description Broad-leaved arrowhead is an emergent aquatic with 4- to 16-inch leaves shaped like exaggerated arrowheads with two long, pointed lobes projecting down. The delicate, white-petaled flowers have yellow centers and grow widely spaced in whorls of three on a leafless stem July to August. Fruits are round and green.

Landscape Use The leaves of broad-leaved arrowhead are extremely variable in width and shape, and a mass planting offers an interesting study in texture. It is not well suited to small pools because it spreads rapidly. Waterfowl, snapping turtles, and muskrats eat arrowhead fruits and tubers.

Site Requires a fertile, heavy, loamy soil at water edges or a submerged soil in full sun or light shade.

Culture Arrowhead grows from a potato-like tuber in water 2 to 3 inches deep, or in the moist soil at edges of ponds or lakes. Plants will eventually multiply and form a large colony. Restrict growth by planting in nursery containers.

Good Companions Grow arrowhead with pickerelweed and American white water lily, two other native emergent aquatics.

Sagittaria latifolia

Sanguinaria canadensis
Bloodroot
Zone 3

Sanguinaria canadensis

Native Habitat Rich, moist woods and floodplains throughout most of Minnesota.

Height 4 to 10 inches

Description Bloodroot has simply elegant pure white flowers in early spring. They are 1½ inches wide with eight to ten petals and golden stamens on smooth, leafless stalks. The beautiful rounded basal leaves are deeply lobed and wrap around the stalks of young plants, unfurling with age. The first blooms will appear early April on sunny south-facing slopes, but the main bloom begins mid to late April.

Landscape Use The short-lived but showy pure white flowers of bloodroot really say spring has arrived. It makes a great groundcover in shade or woodland gardens because the leaves remain through most of summer and are ornamental in themselves. Tuck small groups here and there in shady borders.

Site Prefers moist, humus-rich soil with direct sunlight in early spring and summer shade but will tolerate drier soils if well mulched.

Culture Improve poor or sandy soil by adding generous amounts of peat moss and compost. Mulch plants with a thin layer of pine needles or shredded leaves. Heavy mulch can lead to stem rot at the soil surface. Plants may go dormant during dry spells. The thickened, elongated rhizome exudes red sap when broken. Self-sown seedlings will appear when conditions are favorable and are usually a welcome sight.

Good Companions Plant bloodroot with other early spring bloomers such as false rue anemone, bellworts, spreading Jacob's ladder, and Virginia bluebells. It also combines nicely with lady fern, spring bulbs, and pulmonarias.

Cultivars 'Flore Pleno', also sold as 'Multiplex', is an exquisite double-flowered cultivar. Zone 3.

Senecio aureus [Packera aurea]
Golden ragwort
Zone 3

Senecio plattensis

Native Habitat Moist woods and swamps throughout most of Minnesota.

Height 1 to 2 feet

Description Golden ragwort is an early blooming perennial with handsome clusters of golden-yellow, daisy-like flowers borne atop 1- to 2-foot scapes in May and June. The heart-shaped, toothed, basal leaves resemble violet leaves and are dark green above and purplish beneath, turning yellow in fall.

Landscape Use The showy blooms of golden ragwort persist for several weeks in spring, and the leaves look nice when the plant isn't flowering. The golden-yellow color is a nice contrast with other spring blooms, which are usually pink or purple shades. It thrives in moist soils and will even do well in wet sites such as bog gardens. It can be grown in groups of three to five in perennial borders or used as a small-scale groundcover in wet areas.

Site Prefers a moist, slightly acidic soil in full sun to part shade.

Culture Amend soil with organic matter before planting to ensure adequate soil moisture. Plants in drier sites should be mulched to help keep the soil moist. This rhizomatous plants will spread slowly when conditions are right.

Good Companions Combine golden ragwort with other late-spring bloomers such as blue phlox, columbine, blue flag, goatsbeard (*Aruncus*), and creeping phlox (*Phlox stolonifera*).

Other Species *S. plattensis* (prairie ragwort) is native to open woods, savannas, and prairies mainly in the southwestern half of Minnesota. It is a self-sowing biennial with similar flowers and heart-shaped basal leaves. It is not quite as ornamental but tolerates drier sites. Zone 3.

Silphium laciniatum
Compass plant
Zone 3

Silphium laciniatum

Native Habitat Dry to moist tall-grass prairies and savannas in southern Minnesota.

Height 3 to 12 feet

Description There is nothing small about compass plant. It has lemon-yellow, sunflower-like flowers up to 5 inches across at the ends of branched, hairy, sticky stems July to September. The 12- to 18-inch, cut-leaf blades orient themselves horizontally in a north-south direction to avoid the intense rays of midday sun.

Landscape Use Compass plant can be used in large perennials borders, but it is best suited to prairie gardens. Try to find a spot where you can enjoy the beautifully cut and lobed basal leaves as well as the flowers. Goldfinches, chickadees, and sparrows like the large seeds.

Site Prefers moist, average soil in full sun but tolerates a wide range of soil moisture and pH. Mature plants are drought tolerant.

Culture Compass plant needs plenty of room, so space new plants 3 to 4 feet apart. New plants take a year or two to become established. The large taproot of mature plants can go down 15 feet, making established clumps difficult to divide or move, so choose a site carefully. Young plants may need protection from deer. Plants will self-sow. If you don't want the seedlings, weed them out while they are still young and easy to pull.

Good Companions Plant compass plants with other tall prairie flowers such as rattlesnake master, purple prairie clover, Culver's root, New England aster, Joe-pye weeds, big bluestem, and Indian grass.

Other Species *S. perfoliatum* var. *perfoliatum* (cup plant) is native to low, open woods and wet prairies. It grows 3 to 8 feet tall with leaves that encircle tall stems, forming cups that collect dew and become a drinking fountain for birds. The 3-inch yellow flowers resemble sunflowers and bloom in branched clusters just above the leaves in late summer. It can take light shade. It has a commanding presence in perennial borders, and it can be used at the edge of woodland gardens or in prairie gardens. Goldfinches love the seeds. Plants may self-sow. Aphids can be a problem during hot, dry periods. Zone 3.

Silphium perfoliatum

Sisyrinchium angustifolium
Blue-eyed grass
Zone 3

Sisyrinchium angustifolium

Native Habitat Prairies, savannas, and open woods in all but the southeastern corner of Minnesota.

Height 10 to 20 inches

Description This iris relative has small, blue to violet-blue flowers with yellow centers that form loose clusters on top of flattened stems mid-May to June, with sporadic blooms throughout the growing season. The basal leaves are narrow and grasslike.

Landscape Use Blue-eyed grass should be planted in groups where it can be seen in the morning when the flowers are open. Use it in rock gardens or at the edges of perennial borders or prairie gardens. The fine, grasslike foliage has great garden value when plants are not in flower.

Site Prefers a moist, average, well-drained soil in full sun but tolerates light shade.

Culture Blue-eyed grass is easy to grow, quickly forming large clumps by spreading and by self-sown seedlings. Divide plants every other year for the best bloom. Moderate fertility and summer moisture will encourage repeat bloom. Plants can be cut back lightly after flowering to prevent seed formation and self-sown seedlings.

Good Companions Plant delicate blue-eyed grass with late-blooming spring bulbs, prairie smoke, bird's-foot violet, and pussytoes.

Cultivars and Other Species 'Lucerne' has larger blue-purple flowers and is less aggressive in the garden. Zone 4.

S. campestre (field blue-eyed grass) is native to open woods, savannas, and prairies mainly in the southwestern half of Minnesota. It has paler blue or even white flowers on 16-inch plants. Culture and use are the same. Zone 3.

Smilacina racemosa
False Solomon's seal
Zone 2

Smilacina racemosa

Smilacina stellata fruits

Native Habitat Fertile woods and rocky, wooded slopes throughout most of Minnesota.

Height 1 to 3 feet

Description False Solomon's seal grows from creeping rhizomes. It has attractive, 5- to 7-inch, egg-shaped, pleated leaves with smooth margins that clasp the stem. Conical plumes of small white flowers develop at the end of an arching stem and rise above the foliage late spring into summer. The showy fruits start out green and take on a red marbling in summer, eventually turning translucent red.

Landscape Use Use this shade-lover in woodland or shade gardens in clumps of one to three plants. It can also be used as groundcover under trees. Be sure to place it where you can enjoy the interesting berries that appear in mid to late summer and persist into winter.

Site Requires a moist, rich, acidic soil in light shade to full shade; best bloom comes with some sunlight.

Culture False Solomon's seal is easily grown when the soil conditions are right. If necessary, amend soil with organic matter and acidify before planting. Mulch plants with a thin layer of pine needles or shredded leaves. Groundcover plants should be spaced 12 inches on centers. Plants spread, but rarely become invasive. Slugs can be a problem.

Good Companions Grow false Solomon's seal with other shade plants such as baneberries, foamflower (*Tiarella cordifolia*), bloodroot, Virginia bluebells, ferns, hostas, pulmonarias, and epimediums.

Other Species *S. stellata* var. *stellata* (starry false Solomon's seal) is native to open woods, savannas, and prairies throughout Minnesota. It is a more delicate plant with smoother, narrower leaves, arching zigzag stems to 24 inches tall, and terminal, plumelike clusters of white, star-shaped flowers May to June. The alternate, pointed, gray-green leaves that fold along the midrib are a distinctive feature. Use it in woodland gardens with native wildflowers. Zone 2.

Solidago rigida
Stiff goldenrod
Zone 3

Native Habitat Dry to moist prairies and open woodlands throughout most of Minnesota.

Height 3 to 5 feet

Description Stiff goldenrod has wide, flat, or somewhat rounded clusters of gold flowers in early fall. The light green leaves are rough with short, stiff hairs and turn an attractive dusty rose in fall.

Landscape Use Goldenrods are a staple of prairie gardens, but they also adapt well to sunny perennial borders. The flowers attract bees and butterflies, including monarchs, and work well in fresh and dried arrangements.

Site Prefers average to infertile, well-drained soil in full sun.

Culture Keep the soil on the lean side; plants in rich soil often become floppy. Plants grow from clumping or spreading rhizomes that will spread but can easily be kept in control if needed. Goldenrods are wrongly accused of causing hay fever; they are insect-pollinated plants, so their pollen is heavy and sticky rather than wind-borne. Self-sown seedlings can become weedy in small gardens.

Good Companions Plant goldenrods with other late-summer plants such as asters, Joe-pye weeds, blazing stars, purple coneflowers, and grasses.

Other Species *S. flexicaulis* (zigzag goldenrod) is native to woods and woodland edges in all but far northern Minnesota. It grows 2 to 4 feet tall with erect, somewhat zigzagged stems and flowers in small clusters. It prefers richer soil and light shade and is a good candidate for woodland gardens, where fall color is hard to come by. Zone 3.

S. nemoralis (gray goldenrod) is native to dry prairies and open woodland throughout Minnesota. It has 1- to 2-foot, tight clumps of gray-green basal leaves and one-sided, plumed clusters of lemon-yellow flowers on 2-foot stems. Its smaller stature opens it up to use in rock gardens. Zone 2.

S. speciosa (showy goldenrod) is native to prairies, savannas, and open woods mainly in the southwestern half of Minnesota. It is one of the most beautiful goldenrods, growing 2 to 5 feet tall with small yellow flower heads crowded into dense, pyramidal, terminal clusters. Plants form tight clumps of leafy, red-tinged stems. Divide clumps every third year to promote vigorous growth and better flowering. An excellent choice for perennial borders. Zone 3.

Solidago rigida

Solidago flexicaulis

Thalictrum dioicum
Early meadow rue
Zone 3

Native Habitat Moist woods mainly in the eastern half of Minnesota but also farther west.

Height 1 to 3 feet

Description Early meadow rue is grown mainly for its fine-textured foliage and bushy shape. The much-divided, thinly spreading compound leaves are bluish green. Plants are either male or female, with female flowers being insignificant and the greenish male flowers appearing in mid to late April in candelabra-like clusters of fuzzy-petaled flowers with pendant golden stamens that tremble in the slightest breeze.

Landscape Use Treat early meadow rue like a fern in the landscape, using its foliage to blend and soften other plants. It can be used in shade or rock gardens, along stone walls, or in masses in woodland gardens. The delicate flowers will be much more effective if you can get them up to eye level.

Site Prefers a fertile, moist, slightly acidic soil in light shade but tolerates drier soil and deep shade.

Culture Early meadow rue is easily grown in gardens, seldom needing division or weeding out. Leaf miners can ruin the foliage; get rid of plant debris to remove overwintering sites.

Good Companions Grow early meadow rue with showier woodland wildflowers such as bloodroot, violets, and Virginia bluebells, using the foliage to tie the plantings together.

Other Species *T. dasycarpum* (tall meadow rue) is native to wet prairies, wetland margins, and floodplains throughout Minnesota. It grows 4 to 6 feet in height. The stems are often purple and its inflorescence is made up of numerous drooping white flowers. It needs a larger spot with moist soil and can tolerate more sun. Zone 3.

Thalictrum dioicum

Thalictrum dasycarpum

Tradescantia occidentalis var. occidentalis
Western spiderwort
Zone 3

Tradescantia bracteata

Native Habitat Savannas and dry prairies mainly in east-central Minnesota but also farther west.

Height 12 to 18 inches

Description The foliage has long, leafy stems that resemble daylily foliage, especially when first emerging. Leaves are blue green. The three-petaled flowers are flat faced and deep rose-pink to purple. They only last a day and usually close up by midafternoon, but they bloom for most of June and July.

Landscape Use Plant spiderworts in prairie gardens or on the edge of woodland gardens. Their drought and heat tolerance make them good choices for xeriscaping. They can also be grown in perennial borders, but the foliage often looks old and tired by midsummer and plants may go dormant after flowering, so try to place them where they will be masked by other plants.

Site Grows best on lean, well-drained soil in full sun or light shade, but will grow on heavier, richer soils. Plants grown in shade will have fewer flowers.

Culture In fertile soils, plants may become aggressive. They also reseed. Foliage can be cut back to the ground after flowering to keep plants neat and halt seed production.

Good Companions Plant spiderworts with plants that offer late-season interest such as mountain mints, asters, goldenrods, and ferns.

Other Species *T. bracteata* (bracted spiderwort) is native to dry prairies and savannas in the southwestern half of Minnesota. It grows 1 to 2 feet tall and wide with blue-gray foliage and medium blue flowers, spreading gradually to form clumps. Culture and use are the same. Zone 4.

T. ohiensis (Ohio spiderwort) is native to dry, open, sandy habitats in the far southeastern corner of Minnesota. It's a slender, branching plant growing 2 to 4 feet tall with narrow, blue-green leaves and blue-purple flowers with rounded petals. Culture and use are the same. 'Alba' is a white-flowered form. Zone 3.

Trillium grandiflorum
Large-flowered trillium
Zone 3

Native Habitat Fertile, moist woodlands and floodplains mainly in east-central Minnesota.

Height 10 to 18 inches

Description Large-flowered trillium has a single white flower with three waxy petals and six yellow stamens borne on a stalk above a whorl of three roundish leaves on an erect stem. The funnel-shaped flower is 2 to 4 inches wide and turns pink with age. Flower time is mid to late April through May.

Landscape Use Trilliums really say "Minnesota woodland" and are a great addition to any shade or woodland garden, either as specimen groupings or massed. They make a nice under-planting beneath spring-blooming shrubs and trees.

Site Requires rich, well-drained soil in light to dense shade.

Culture Large-flowered trillium adapts surprisingly well to cultivation. The ideal site is a humus-rich sandy loam soil that remains consistently moist where plants receive spring sun and summer shade. If massing, space plants 10 inches on center. It can take as many as seven years to flower when grown from seed. Be sure to only purchase container-grown, nursery-propagated specimens.

Good Companions Set off the pure white flowers with blue phlox, Virginia bluebells, wild geraniums, and woodland ferns.

Cultivars and Other Species 'Flore Pleno' is a rare, fully double-flowered form.

T. cernuum var. *macranthum* (nodding trillium) is native to rich, damp woods throughout Minnesota. It has short, curved flower stalks so the flowers hang beneath the leaves. The nodding white flowers have reflexed petals that reveal rose or maroon anthers. It blooms anywhere from late April to June. Zone 2.

T. nivale (snow trillium) is native to moist forests and bottomlands in southern Minnesota. It is the earliest trillium, starting its bloom in early April, sometimes poking up through snow. Plants are smaller than large-flowered trillium with narrower leaves and flowers. Grow this 6-inch plant where it will be seen up close or it will get lost. It adapts well to cultivation, but it is difficult to locate. It is on Minnesota's special concern list, so only buy nursery-propagated plants. Grow it in moist, humus-rich, limy soil in light to partial shade. Plants usually go dormant after flowering. Zone 4.

Trillium nivale

Trillium grandiflorum

Uvularia grandiflora
Large-flowered bellwort
Zone 3

Uvularia grandiflora

Native Habitat Deciduous woods and bottomlands usually in limy soils throughout most of Minnesota.

Height 12 to 16 inches

Description Large-flowered bellwort has 1- to 2-inch, lemon-yellow flowers that dangle bell-like from downturned stems late April to mid-May. The alternate, long, sea green leaves have smooth margins and downy white undersides and clasp the stem at their bases. The foliage expands after flowering to create a soothing, soft green groundcover.

Landscape Use The bright yellow color is a real attention-getter in early spring woodland gardens. Place clumps here and there for interest. It is a good choice next to foundations on the north sides of buildings, as it can take the higher pH soil conditions often found next to concrete blocks. Plants compete well with the roots of trees, and the long-lasting, attractive foliage is an added bonus.

Site Prefers a moist, average, neutral to limy soil in spring sun and summer shade.

Culture An annual dusting of limestone may be needed in some sites to keep the soil pH in line.

Good Companions Large-flowered bellwort grows well with almost any early flowering woodland plant, but it is especially attractive with blue flowers such as Virginia bluebells and woodland phlox. In shade gardens, grow it with epimediums, foamflower, and pulmonarias.

Other Species *U. sessilifolia* (pale bellwort, wild oats) is native to woods and thickets mainly in northeastern Minnesota but also farther west and south. It's smaller, growing 4 to 12 inches tall, and more delicate and lighter yellow in flower. Culture and use are the same, but it prefers a more acidic soil. 'Variegata' is a rare variegated form for collectors. Zone 3.

Uvularia sessilifolia

Verbena stricta
Hoary vervain
Zone 3

Verbena stricta

Native Habitat Dry prairies and savannas in the southern half of Minnesota.

Height 1 to 3 feet

Description Hoary vervain has erect branches holding strongly vertical spikes of blue-violet flowers in summer and early fall. Bloom starts at the bottom of the flower spike and works its way up, resulting in a long bloom time. The leaves are wedge-shaped and hairy.

Landscape Use The blue flowers are easily combined with gold and yellow late-summer bloomers in perennial borders, where they add a soft vertical accent. Vervains should be a part of any prairie garden, and they make good cut flowers.

Site Tolerates a wide variety of soil types in full sun.

Culture Hoary vervain adapts well to garden use. Self-sown seedlings occur but are rarely a nuisance.

Good Companions Grow hoary vervain with wild bergamot, phloxes, coneflowers, and grasses.

Other Species and Cultivars

V. hastata (blue vervain) is native to wet prairies, wetland margins, and marshes throughout Minnesota. The 3- to 5-foot erect branches hold strongly vertical candelabra-like spikes of small, dark blue flowers in summer and early fall. The leaves are long and narrow-toothed. Plants are a little too coarse for specimen use, but the dark blue flowers are easily combined with gold and yellow late-summer bloomers and grasses in perennial borders and prairie gardens, where they add a strong vertical accent. It can also be grown in bog gardens and alongside streams and ponds. It prefers rich, evenly moist to wet soil in full sun or light shade. 'Alba' is a hard-to-locate cultivar with pure white flowers. Zone 3.

Verbena hastata

119

Vernonia fasciculata
Bunched ironweed
Zone 3

Vernonia fasciculata

Native Habitat Low, wet places and prairies mainly in the southern half of Minnesota but also farther north.

Height 3 to 6 feet

Description Bunched ironweed is a tall, clumping plant with medium green leaves and clumpy panicles of reddish purple flowers in late summer.

Landscape Use Bunched ironweed can be used toward the back of large perennial borders where it offers a strong vertical presence. It is also good for prairie gardens and screening. The flowers are good for cutting.

Site Prefers moist, well-drained soil in full sun but once established it will tolerate drier soils and drought.

Culture Although native to moist soils, this perennial adapts well to garden settings as long as it is not drought stressed. Mulch plants to keep soil moist. Pinch back stems in late May to keep plants more compact. Unpinched plants may need staking or support of some kind.

Good Companions Plant bunched ironweed with other late-summer bloomers such as goldenrods, oxeye, Joe-pye weeds, and boltonia.

Veronicastrum virginicum
Culver's root
Zone 3

Veronicastrum virginicum

Native Habitat Prairies, floodplains, and woodland edges in all but far northeastern Minnesota.

Height 3 to 6 feet

Description Culver's root is a strong architectural plant. The dark green, whorled leaves give the plant a horizontal effect that contrasts with the strong vertical spires of white or pale lavender, candelabra-like flowers that start blooming in midsummer and continue into early September.

Landscape Use This durable plant can be used in the middle or back of a perennial bed, where it offers a strong upright accent. It tolerates wet soil well. The flowers are good for cutting.

Site Prefers a moist to wet, rich soil and full sun to light shade, but grows well in ordinary garden soil that has been enriched with ample organic matter.

Culture Work generous amounts of organic matter into the soil before planting. Mulch plants and give them an annual application of compost to keep soil rich and moist. Culver's root forms clumps as it ages but is not overly aggressive. Water plants during dry spells.

Good Companions The long bloom time and tall stature of Culver's root make it a good backdrop for many border plants, including blazing stars, lilies, oxeye, monardas, milkweeds, rattlesnake master, goldenrods, and asters.

Cultivars 'Apollo' has lavender flowers. 'Roseum' has pale rose-pink flowers. Zone 3.

Viola pedata
Bird's-foot violet
Zone 3

Native Habitat Prairies and woodland openings in southeastern Minnesota.
Height 4 to 8 inches
Description Bird's-foot violet has bluish purple flowers up to 1½ inches wide, bilaterally symmetrical with five petals and orange centers. They start blooming in late April. The leaves, which are shaped like bird's feet, rise directly from the roots.
Landscape Use Use this charming plant as groundcover on sandy banks or in large rock gardens. It also adapts well to container culture and is pretty in trough gardens. Plants may rebloom sporadically in summer and fall.
Site Prefers sandy, acidic, well-drained soil in full to partial sun.
Culture Bird's-foot violet is a bit more finicky than other violets, but worth the trouble. It will rot if the soil is not well drained, and it's not a strong competitor, so keep it free of weeds and other aggressive plants. Maintain soil acidity with a mulch of pine needles or shredded oak leaves. Plants may be short-lived but seedlings are plentiful. Rabbits can be a problem.
Good Companions Plant bird's-foot violet with other delicate early bloomers such as pasque flower, prairie smoke, blue-eyed grass, and spring bulbs. It really stands out against the brown grasses when it blooms in prairie gardens in spring.
Other Species *V. pubescens* (yellow violet) is native to deciduous woods and floodplains throughout Minnesota. It is a robust plant growing 6 to 16 inches tall with lemon-yellow flowers. Plants have hairy, heart-shaped leaves and form dense clumps in rich, evenly moist soil in light to full shade. It blooms a little later than bird's-foot violet. Use it in woodland or shade gardens. Zone 3.

V. sororia (common blue violet) is native to woodlands and clearings throughout Minnesota. It grows about 10 inches tall with deep purple-blue flowers and makes a lush green groundcover where few plants will grow. It likes moist soil in sun or shade but tolerates a wide variety of conditions. It is a prolific self-sower and can become weedy. Zone 3.

Viola pedata

Waldsteinia fragarioides var. fragarioides
Barren strawberry
Zone 4

Native Habitat Open woods and savannas in northeastern Minnesota.
Height 3 to 6 inches
Description Barren strawberry is a rhizomatous perennial somewhat resembling cultivated strawberry plants. The basal leaves have three leaflets and the flowering stems bear several small yellow flowers April to May followed by tiny inedible fruits.
Landscape Use This unique-looking groundcover forms a dense mat where conditions are favorable. It leafs out early and can tolerate the dry shade under large trees or on slopes. It can be used in large rock gardens, stone walls, or between pavers in walkways.
Site Prefers a moist, acidic soil in full sun to light shade but tolerates drier sites.
Culture If necessary, before planting, acidify soil and improve drainage with organic matter and sand. Plant crowns even with the soil surface as you would plant strawberries. Plants are fairly drought tolerant once established but require regular watering their first year. Barren strawberry is on Minnesota's special concern list, so only purchase nursery-propagated plants.
Good Companions Barren strawberry makes a good groundcover under azaleas and other shrubs and large trees.

Waldsteinia fragarioides

Zizia aurea
Golden alexanders
Zone 3

Native Habitat Open woods, flood-plains, and moist prairies throughout Minnesota.

Height 1 to 3 feet

Description This erect, bushy perennial has umbels of small yellow flowers that resemble Queen Anne's lace from May to June. The compound leaves have elongated leaflets and a bluish cast.

Landscape Use The flattened heads of the yellow flowers are a bright accent in spring gardens and the foliage looks nice all summer. Use it in prairie gardens, open woodland gardens, or perennial borders, where it will weave its way through other plants. It spreads rapidly and should not be planted where space is limited. Plants are food sources for several butterfly larvae.

Site Prefers light shade or full sun and moist, well-drained soil.

Culture Once established, alexanders are carefree plants that tolerate summer dryness. Individual plants may be short-lived, but they will self-sow to maintain the population.

Zizia aurea

Good Companions The cheery yellow flowers combine nicely with prairie smoke, blue phlox, butterfly weed, wild lupine, columbines, and wild geraniums.

Other Species *Z. aptera* (heart-leaved alexanders) has a similar native range. Flowers are similar but smaller, and the 1- to 2-foot flower stalks arise from rosettes of attractive, leathery basal leaves. It is not as invasive as golden alexanders and can be used in more-formal landscape situations. Unfortunately, it is difficult to locate. Zone 3.

Additional Native Flowers and Groundcovers

Anaphalis margaritacea (pearly ever-lasting) is native to dry-soil areas mainly in the northeastern quarter of Minnesota. This 1- to 3-foot bushy plant has flat-topped clusters of small, silvery-white, daisylike flowers July into September. The narrow leaves are grayish above, woolly white underneath, offering a nice contrast in mixed borders. They are long-lasting dried flowers when the flowers are picked before they are completely open. The flowers are attractive to butterflies. Pearly everlasting is tolerant of tough, dry conditions and can be grown in poor to average soils in full sun. Zone 2.

Astragalus canadensis (Canada milk vetch) is native to prairies and open woodlands throughout most of Minnesota. This robust clump perennial grows 1 to 4 feet tall and has pinnately compound leaves with many leaflets. The elongated, loose clusters of interesting flowers are cream colored and appear in July and August. They are attractive to hummingbirds and can be used in cut bouquets. Grow Canada milk vetch in full sun to part shade in perennial borders and prairie gardens. It does not transplant well, so choose a site carefully. Zone 2.

Calla palustris (wild calla) is native to bogs, swamps, and ponds in northeastern Minnesota. It grows 12 inches tall and has a broad, white, 2-inch spathe partially surrounding a 1-inch spadix bearing tiny yellow flowers May to August. The long-stalked basal leaves are dark green, glossy, heart-shaped, and 2 to 6 inches long. Plant it at the edge of streams or ponds in submerged soil in full sun where it can form colonies. It will extend the season of color after marsh marigolds go dormant mid-summer. Zone 2.

Delphinium carolinianum var. virescens (prairie larkspur) is native to dry, gravel prairies, savannas, and woodland edges in the southwestern half of Minnesota. It is a slender, sparsely branched plant growing 3 to 5 feet tall with slender spikes of white to pale blue flowers in midsummer. Plant it in sandy, well-drained soil in full sun or light shade in prairie gardens. Zone 3.

Epilobium angustifolium (fireweed) is native to moist, open areas in northeastern Minnesota. It grows 3 to 6 feet tall with spiked clusters of flamboyant magenta flowers that bloom from the bottom up, starting in July and continuing into late August. It is especially abundant in an area two to three years after fire, where it can grow into dense groundcover. Fireweed is a little too aggressive for most landscape situations, but it can be used in naturalized plantings and for erosion control. The flowers are attractive to hummingbirds and bees. It requires a moist, well-drained soil in full sun. Zone 2.

Astragalus canadensis

Equisetum hyemale var. affine (tall scouring rush) is native to damp places near streams throughout Minnesota, except the far southeastern corner. It is an interesting plant growing 20 to 36 inches tall with evergreen, unbranched, jointed stems. It is an instant focal point in Japanese-style gardens or near garden pools. Grow it in moist, slightly acidic soil in partial shade to almost full sunlight. It can become invasive; plant it in a buried nursery container to keep rhizomes contained. Zone 2.

Fragaria virginiana (wild strawberry) is native to open woods, savannas, and prairies throughout Minnesota. It has 1-inch-wide white flowers that peek out between deep green, three-part leaves in

May followed by small bright red, juicy fruits that grow up to ¾ inch in size. The strawberries are edible and tasty—if you can get to them before the wildlife. The 6-inch plants spread by runners and a mass planting makes an interesting groundcover in sun to partial sun in average to dry soil. It can also be grown in a rock wall or a rock garden. Zone 2.

Galium boreale (northern bedstraw) is native to prairies and open woodlands throughout Minnesota. It grows about 12 inches tall from creeping roots and forms large patches, weaving its way among other plants. It is easy to grow in any soil in sun or partial shade and it is drought tolerant once established. Use it to create a lacy backdrop for darker-colored prairie plants such as butterfly weed and bird's-foot coreopsis. Zone 2.

Gaura coccinea (scarlet gaura) is native to dry prairies and savannas along the western border of Minnesota. The 2-foot stems are densely covered with narrow leaves and ¼-inch, rose-pink to red flowers crowd at the tips of stems. Plant it in average to rich, well-drained soil in full sun in prairie gardens or perennial borders. Zone 3.

G. biennis (biennial gaura) is native to mesic prairies, savannas, or grassy openings in deciduous woods in the far southeastern corner of Minnesota. It is a self-sowing annual or biennial that often shows up in wildflower seed mixes. Zone 4.

Galium boreale

Helianthus annuus (**common sunflower**) is native to dry prairies throughout Minnesota. It is an annual growing up to 10 feet tall with one or more large terminal sunflowers on rough stems. Bloom time is July to October. Use it in children's gardens, to attract birds, and for screening. Plant it in average, well-drained soil in full sun. Many cultivars are available.

H. giganteus (giant sunflower) is native to swamps and wet prairies throughout most of Minnesota. It is a perennial growing 6 to 10 feet tall with loose, branching panicles of 2½-inch flowers. Zone 3.

H. maximiliani (Maximilian sunflower) is native to dry prairies throughout most of Minnesota. This perennial grows 6 to 8 feet tall with 12-inch leaves and short-stalked, yellow flowers. Both perennial sunflowers are coarse plants with limited garden use, but they can be used for screening and in prairie gardens. Zone 3.

Heuchera richardsonii (**alumroot**) is native to prairies, savannas, and open woods throughout Minnesota. It grows from a 12-inch-tall basal clump of heart-shaped, long-petioled, green leaves that show some white mottling or a purple blush when young, maturing to a more-uniform green. Tiny, greenish, bell-shaped flowers in open, airy panicles are borne on slender, wiry stems extending well above the mound of leaves, typically to a height of 2 feet in spring to early summer, long enough to emerge above prairie grasses. Alumroot's attractive foliage and airy flower panicles provide color and contrast to rock gardens, perennial borders, or open woodland gardens. It tolerates dry locations and is a good edging plant. It can be massed to form an attractive groundcover. Zone 2.

Heuchera richardsonii

Hydrastis canadensis (**golden seal**) is native in the southeastern corner of Minnesota in deciduous forests usually in sheltered ravines or on talus slopes. It grows 6 to 16 inches tall with attractive maple-like leaves and single white flowers with yellow stamens in late April and May, followed by red fruits. It is an endangered plant in Minnesota and in most states; growing nursery-propagated specimens will help perpetuate the species. Use it in woodland gardens with spring beauties, hepatica, and other spring-bloomers. Zone 4.

Hydrophyllum virginianum

Hydrophyllum virginianum (**Virginia waterleaf**) is native to moist to wet woods and openings in most of Minnesota. It has showy, silvery gray "watermarked" leaves that provide early interest in woodland gardens. The pale to deep violet flowers appear in airy, rounded clusters in mid-May. Plants are easy to grow but can become weedy. Zone 3.

Lespedeza capitata (**round-headed bush clover**) is native to dry prairies and savannas mainly in southern Minnesota. It grows 30 to 48 inches tall on sturdy stalks. Inconspicuous whitish flowers emerge in summer, but the main attraction is the bronze seed heads that provide fall and winter color and are an important food supply for birds. Plant it in a dry, well-drained sandy soil in full sun or very light shade in prairie gardens. Zone 4.

Linnaea borealis (**twinflower**) is native to moist, cool, northern woodlands and bogs in the northeastern quarter of Minnesota. It is a creeping evergreen growing 3 to 6 inches tall with stems that root as they spread. Each upright branch bears a pair of nodding white or pinkish, bell-shaped flowers from June to August. The rounded, light green leaves are finely toothed at their tips and occur in pairs along the trailing stems. Twinflower is a must if you are trying to recreate the northern coniferous forest in your landscape. Plant it near the edge of paths where you can see it up close. This dainty little groundcover looks very nice in mossy habitats and growing over rotting stumps and logs. It requires a moist, high-humus, acidic, peaty soil in partial shade. Plants will form dense colonies in time but do not become invasive. Zone 2.

Lysimachia ciliata (**fringed loosestrife**) is native along streams, usually in moist, shaded places but sometimes in full sun, throughout Minnesota. It grows 2 to 3 feet tall and forms a sprawling mass. The pretty yellow flowers are arranged in whorls and are borne on slender stalks arising from leaf axils June to September. Use it for naturalizing in low, moist areas in light shade; it is aggressive and may overcrowd nearby plants. 'Atropurpurea' and 'Purpurea' have purple-bronze foliage and are supposedly less invasive. Zone 3.

Lysimachia ciliata

Mitchella repens (partridgeberry) is native to rich, moist woodlands, pinelands, and clearings mainly in east-central Minnesota but also farther north and south. It is a low, creeping evergreen 1 to 3 inches tall that forms large mats. It has paired, white, tubular flowers with four spreading lobes in May. Each pair of flowers forms one bright red berry in fall. Leaves are opposite, round, and shiny green with a distinct white vein down the center. Use this dainty groundcover in shade gardens, where it will weave among other shade plants. It makes a nice underplanting for acid-loving shrubs such as azaleas. The wintergreen-scented berries are showy all winter if not eaten by birds. Mulch with a thin layer of pine needles, and water during dry periods. Do not allow this low-growing plant to be smothered by fallen leaves. Zone 3.

Napaea dioica (glade mallow) is native to moist, sunny spots along rivers or in partial shade near adjacent woodland borders in far southeastern Minnesota. It is a large, robust plant that can grow to 6 feet tall with several flowering stems. The fragrant, white, five-petaled flowers bloom for about a month in midsummer. It prefers a well-drained, humus-rich soil in full or partial sun. It is intolerant of drying out, the leaves rapidly wilting, and it is suitable for bog gardens. The tall flowering stems are vulnerable to wind damage. It is on the Minnesota threatened species list, so only purchase nursery-propagated specimens. Zone 4.

Parthenium integrifolium (wild quinine) is native to open woodlands in the far southeastern corner of Minnesota. It grows 2 to 4 feet tall with upright stems holding leathery, dark green leaves. It has pure white flowers in dense clusters from midsummer into fall. They are fragrant and make excellent cut flowers. It grows in a wide range of soils and is drought tolerant once established. Plant wild quinine in full sun or part shade in prairie gardens and perennial borders. It is on Minnesota's endangered species list, so only purchase nursery-propagated plants. Zone 3.

Potentilla tridentata (three-toothed cinquefoil) is native to acidic, well-drained soils mainly in northeastern Minnesota. It forms a 3- to 8-inch mat of dark green, strawberry-like leaves that turn a nice wine-red color in fall. The small white flowers are numerous, starting late May and continuing into summer. It needs a cool, acidic, well-drained, sandy or even gravelly soil in sun to part shade. Use it in rock gardens or for edging, where it will spread but not become too weedy. Zone 2.

P. arguta (tall cinquefoil) is native to dry soils throughout Minnesota. It grows 1 to 3 feet tall with pale yellow flowers all summer, tolerating dry conditions and full sun in rock gardens and perennial beds. Zone 2.

Streptopus roseus var. longipes (rose twisted stalk) is native to rich woods and stream banks in far northeastern Minnesota. It has 12- to 30-inch succulent stems clothed in clasping, deeply veined, succulent leaves. Small rose-pink bells hang from the leaf axils in spring. Plant small groups or large colonies in cool shade in moist but not wet sites in rich, acidic soil. Zone 3.

Trientalis borealis (starflower) is native to rich, moist, acidic woods and bogs mainly in the northeastern quarter of Minnesota. This delicate ephemeral grows 4 to 8 inches tall and spreads by thin rhizomes with fibrous roots. The one to three, but usually two, white flowers have golden anthers and arise on slender stalks from a whorl of five to nine canoe-shaped leaves. It blooms mid-May into June. Starflower is a good groundcover in moist, rich, acidic soils, spreading rapidly but not too aggressively. Plants go dormant midsummer, so plant them with ferns and other summer-foliage plants. Mulch with sphagnum moss or pine needles to maintain soil acidity, and grow in full or part shade. Zone 2.

Trientalis borealis

Grasses and Sedges

Grasses are staples of the prairie ecosystem, but many are ornamental enough to be used in other areas of the landscape as well. They are easy to grow and care for, and they wave in the slightest breeze, bringing a sense of movement to a landscape. They have four-season interest, but are especially valuable in fall and winter, when they combine beautifully with evergreens and fruiting shrubs. Taller types can be used similar to shrubs, as background plants, as specimens, or even as a hedge. Smaller types work well with annuals and perennials in herbaceous borders.

Grasses have a pure, abstract quality that blends well with modern architecture, and their shapes, colors, and textures contrast nicely with wood, stone, and other hard structural surfaces. Some people choose to create entire gardens of grasses. Beyond their landscape value, many are valuable sources of food and cover for birds, and the mature seed heads are prized for dried arrangements.

Being wind-pollinated, grasses don't need brightly colored flowers to attract insects. They release their pollen into the air to float from flower head to flower head on gentle breezes. That doesn't mean their flowers aren't attractive, however. Flowers can be lacy panicles, stiff brushes, or waving plumes, often with beautiful fall color. Most grasses flower from August into fall, but some flower earlier. To make it easier to see the flowers, plant grasses against dark fences or backgrounds of evergreens.

Leaves of grasses are lance shaped with parallel veins and come in many shades of green. Plants have round, hollow stems and branching roots, never a taproot. Grasses are classified as mounded, upright, arching, open, irregular, and combinations of these forms. Most grow from clumps and are easy to use in the landscape, but some spread by creeping rhizomes (underground stems)

Native grasses are essential to prairie gardens, but many can also be used to add interest to other areas of the landscape in autumn and winter. Here *Panicum virgatum* 'Rostrahlbusch' grows with *Picea pungens* 'Globosa' (Colorado spruce) and *Hylotelephium* 'Autumn Joy' (sedum).

or stolons (horizontal stems that creep just above or just below ground and root at nodes). These spreading types are best used in prairie gardens, for erosion control, or groundcover.

Sedges are grasslike plants, but are not true grasses. They can be distinguished from grasses by their solid, three-angled flower stems (grasses have round, hollow flower stems). Most sedges form dense, compact clumps of bright green foliage.

Planting

Some native grasses are only available as seeds and often in mixtures used for seeding prairies or other restoration projects. Many of the types better suited to landscape use are available in containers, and this is the easiest way to plant them. Plant after all danger of frost has passed in spring, spacing plants according to their final size. Smaller grasses can be planted as close as 1 foot apart, with taller types requiring 3 to 5 feet between plants. Water well after planting and as needed the first year until plants are established. A summer mulch will help conserve water and keep weeds down. Place 2 inches of compost, shredded leaves, dried grass clippings, or pine needles around plants in late spring.

Most grasses grow best in well-drained soil that is not overly fertile. In all cases, the addition of organic matter to the soil increases plant health. If possible, before planting mix in compost, rotted manure, or peat moss. Most grasses flower and grow best in full sun, but some tolerate light shade, and a few prefer shade. Most are tolerant of wind and can be used in open, exposed areas where other plants would be damaged. It is important to remove all grassy weeds before planting, since it is difficult to weed out grass weeds from grass plants. If not eliminated, weedy perennial grasses like quack grass and canary reed grass will overwhelm desirable plants.

Care

Grasses require little maintenance once established. To help ensure survival the first winter after planting, cover the soil with weed-free straw or leaves after the ground has frozen. Remove mulch before growth begins in spring. On established plants, leave the foliage so you and the birds can enjoy the flowers in winter. Cut back plants in late winter before new growth begins. If you only have a few plants, use a hand pruner or hedge shears. A string trimmer can be used if you have a large number of grasses to cut back. If practical, burning is also effective.

To look their best, grass plants need dividing at some point, some more frequently than others. Early spring is best. If the clump is crowding its neighbors or looks dead in the middle, remove the center and replant the smaller outside clumps.

Possible Problems

Grasses are generally disease and insect free. Poorly drained soils may lead to crown rot on some plants. To reduce problems, choose the right grass for the site or improve the soil before planting.

Andropogon gerardii
Big bluestem
Zone 3

Andropogon gerardii

Native Habitat Prairies and woodland openings throughout Minnesota.

Height 4 to 8 feet

Description Big bluestem is a clump-forming grass with blue-green stems topped with fingerlike racemes of fruiting clusters. The flower clusters, which resemble upside-down turkey feet, really stand out in late summer. The fall foliage color is an attractive bronzy reddish brown.

Landscape Use Although attractive, big bluestem gets too large for many landscape situations. Use single plants in prairie gardens or large mixed borders. It is great for massing in natural-istic plantings, where birds can enjoy the seeds.

Site Prefers a moist, fertile, well-drained soil in full sun to part shade but is adaptable to soil type.

Culture This warm-season grass is easy to grow. Space plants 1 to 2 feet apart. Mow or cut back in spring. Big bluestem is long-lived, so choose a site carefully.

Good Companions Combine this large grass with other tall prairie plants such as asters, goldenrods, Joe-pye weeds, and Indian grass, or grow it as a backdrop for butterfly weed, milkweeds, coneflowers, and blazing stars.

Bouteloua curtipendula
Side-oats grama
Zone 3

Bouteloua curtipendula

Native Habitat Prairies and open woodlands in the southwestern half of Minnesota.

Height 2 to 3 feet

Description Side-oats grama gets its name from the way the flowers are arranged on one side of the stem. The purplish spikes of oatlike flowers appear in midsummer, hanging downward from the stalks, and the seed heads remain attractive into fall. It produces short rhizomes and tends to a bunch-type growth. This warm-season grass has gray-green summer foliage changing to straw color in fall.

Landscape Use This grass is often overlooked for a specimen, but the unique flowers serve as a striking focal point in prairie gardens or sunny rock gardens. Birds eat the seeds in winter. This sod-forming grass can take occasional mowing and is sometimes part of native lawn mixtures.

Site Prefers an average to dry, well-drained garden soil in full sun.

Culture Space plants 1 foot apart. Mow or cut back in spring. Plants spread relatively slowly, but plant it in a buried nursery container if you need to restrict root growth.

Good Companions Plant side-oats grama with other dry-soil prairie plants or massed as a groundcover.

Other Species *B. gracilis* (blue grama) is native to limited-rainfall areas mainly in the southwestern half of Minnesota. It grows 1 to 2 feet tall in flower, but the foliage is less than a foot tall. It has fine-textured leaves and attractive flower spikes in late summer. It is drought tolerant and can grow in dry, gravelly soil. Use it in dry prairies and rock gardens. Zone 3.

B. hirsuta (hairy grama) is native to sandy and rocky soils in the southwestern half of Minnesota. It grows 2 to 3 feet tall and has attractive purple and orange flowers. It is excellent for dry soils as well as well-drained loams and clay and is effective when planted with other short prairie grasses and forbs. Zone 3.

Carex pensylvanica
Pennsylvania sedge
Zone 3

Native Habitat Moist woods throughout most of Minnesota.

Height 8 to 10 inches

Description This grasslike plant grows in cute but persistent tufts on the forest floor. Its long rhizomes allow it to spread out and colonize nearby open areas, and it often appears in pure stands in woodlands. This cool-season plant provides some of the earliest green color in spring.

Landscape Use Use individual clumps or small groupings to soften stones in woodland gardens or shady rock gardens where Pennsylvania sedge's uncommon grasslike form will add interest. It is sometimes planted as a no-mow groundcover under maples, basswoods, or pines.

Site Grows on well-drained sites, with soils ranging from loam to sand. Some soil types are slightly acidic and relatively infertile. Does well in shade or sun.

Culture Pennsylvania sedge is adaptable to a wide variety of conditions. It spreads, but is not overly aggressive. It does not appear to be bothered by deer.

Good Companions Pennsylvania sedge looks nice with almost any woodland plant and is especially effective planted with spring ephemerals.

Other Species *C. grayi* (Gray's sedge) is native to moist soils in southeastern Minnesota. This semi-evergreen sedge grows 24 inches tall with starlike seed heads that are interesting in dried arrangements. It is tolerant of wet soil and partial shade. Zone 4.

C. muskingumensis (palm sedge) is native along rivers and in low woods in far southeastern Minnesota. It grows 2 to 3 feet tall with wide, straplike, light green leaves that resemble palm fronds. Grow it in moist, fertile soil in partial to full shade. The creeping rhizomes spread slowly to form an effective groundcover. 'Wachtposten' is more upright and tolerates drier sites. Zone 4.

C. plantaginea (plantain-leaved sedge) reaches Minnesota at the extreme western edge of its range in moist woods. It has broad evergreen leaves that are reddish at the base. Use it to soften stones along shady water gardens. It is on Minnesota's endangered species list, so only buy nursery-propagated plants. Zone 3.

C. stricta (tussock sedge) is native to wetlands throughout most of Minnesota. It forms a dense, 2-foot clump of long, slender leaves that arch outward, creating a symmetrical fountain-like effect. As the plants develop, they form vertical-sided columns on which they grow, raising them above their surroundings and further adding to their distinctive appearance. Zone 3.

Carex pensylvanica

Carex muskingumensis

Deschampsia caespitosa
Tufted hair grass
Zone 4

Native Habitat Fertile wetlands or sandy lakeshores in the northern half of Minnesota.

Height 3 to 4 feet in flower

Description Tufted hair grass is a well-mannered bunch grass with dark green stems and rolled leaves that give the plants a stiff, wiry appearance. During summer it produces large, open panicles of glistening silver-tinted flower heads with a cloudlike quality. The tight basal tufts grow about 1 foot tall and spread 2 feet wide.

Landscape Use A tolerance for wet soils makes this plant useful for planting near water; it can even be used in bog gardens. It is good for naturalizing, as the billowy masses of fine-textured flower stalks become cloud-like. A dark background will set off the delicate flowers.

Site Tufted hair grass grows best on moist soils in full sun but it does tolerate part shade. Avoid sunny, droughty conditions.

Culture This cool-season grass needs regular water; it will turn brown if allowed to dry out. Garden plants grown in drier soil should be mulched and given supplemental water as needed. It tolerates a fair amount of shade, but flowering will be reduced. Plants often self-sow.

Good Companions Plant tufted hair grass with other moisture-lovers such as cardinal flower, Joe-pye weeds, blue flags, and obedient plant.

Cultivars 'Bronze Veil' ['Bronzeschleier']

Deschampsia caespitosa

has flower heads with distinct bronze-yellow tones when they open, fading to amber as they mature. 'Northern Lights' has cream and green variegated leaves with a pinkish blush on new growth. 'Scotland' ['Schottland'] has dark green leaves and airy, yellow-green flower panicles that fade to light buff. Zone 4.

Elymus canadensis
Nodding wild rye
Zone 3

Native Habitat Mesic prairies and open woodlands throughout most of Minnesota.

Height 3 to 6 feet

Description This cool-season bunching grass has dense, silvery leaves and large, nodding, golden seed plumes July to August that resemble cultivated rye seeds.

Landscape Use This fast-growing prairie grass is not only attractive in prairie gardens, it is a good grass to plant on bare ground, as it looks nice, keeps out weeds, and reduces erosion. Birds eat the seeds.

Site Nodding wild rye prefers a well-drained soil with plenty of moisture in full sun but tolerates full shade and dry soils.

Culture Nodding wild rye grows on an incredible range of soils, including bare sand, gravel, clay, and even damp soils. It also serves as an excellent native nurse crop for prairie seedlings. Plant it at a rate of 2 to 3 pounds per acre with a prairie seed mix. It will mature in the first or second year, ahead of the longer-lived prairie grasses and flowers.

Good Companions Nodding wild rye combines nicely with any prairie plants.

Other Species *E. hystrix* (bottlebrush grass) is native to oak savannas throughout much of Minnesota. It is a tall, graceful grass with large, showy, bottlebrush seed heads that catch the sunlight, even in light shade. Use it in prairie gardens and restorations. Zone 3.

E. virginicus (Virginia wild rye) is native to oak savannas throughout Minnesota. It has green leaves, and the seed heads are upright rather than nodding and they appear earlier. It is also shorter (to 5 feet), less drought tolerant, and more shade tolerant. It

Elymus canadensis

does particularly well in slightly moist soils along woodland streams and floodplains, where it will readily re-seed. It is excellent for wooded openings and forest edges, especially on cooler north- and east-facing forest edges. Zone 3.

Panicum virgatum
Switch grass
Zone 4

Panicum virgatum

Native Habitat Mesic to wet prairies and along ponds and streams throughout most of Minnesota.

Height 3 to 6 feet

Description This warm-season grass forms handsome upright clumps of medium-textured foliage. Airy conical flower heads appear in late summer and are green with a pinkish cast, turning yellow to golden brown in fall.

Landscape Use Switch grass is one of the best native grasses for landscape use. It can be used in mixed borders, as a screen, or in natural gardens. The open flower panicles look best when viewed against a dark background. The dense foliage stands up well in winter and provides excellent cover for wildlife.

Site Any average garden soil in full sun.

Culture Allow 2 to 3 feet between plants, as clumps become large. Switch grass tolerates poor conditions, including poor drainage and occasional flooding. Plants are fairly slow to spread, but division will be needed to keep plants under control in gardens. Self-sown seedlings will appear.

Good Companions Plant switch grass with other late-summer prairie plants such as asters, goldenrods, coneflowers, and boltonia. It looks nice with an evergreen background in winter.

Cultivars 'Heavy Metal' has metallic blue foliage that turns yellow in fall. Zone 4. 'Rotstrahlbusch' has good red fall color. Zone 4. 'Shenandoah' has reddish purple foliage color by midsummer and a distinct reddish cast to the 3-inch flower heads. Zone 4.

Schizachyrium scoparium var. *scoparium*
Little bluestem
Zone 3

Native Habitat Well-drained soils of prairies and savannas in all but far northeastern Minnesota.

Height 2 to 3 feet

Description This attractive clump former has light green to blue foliage in summer, turning golden to reddish brown in fall. The slender stems hold attractive silvery white seed heads. The fluffy seed heads and crimson-colored foliage are extremely showy in the fall landscape.

Landscape Use Little bluestem is among the best native grasses for fall color, and its small size makes it easy to use in most landscapes. Plant it in mixed borders and prairie gardens, along walkways, and in foundation plantings.

Site Prefers well-drained sand or loam in full sun, but will grow in rocky soils and part shade. Not recommended for heavy clay or damp soils.

Culture Little bluestem is a warm-season grass and is slow to emerge in spring. Burn or mow clumps in late winter. It will not do well on heavy soils that hold moisture. It is sometimes sold under the outdated name of *Andropogon scoparius*.

Good Companions The blue-green foliage provides a great backdrop for summer prairie flowers as well as perennials in mixed borders. Plant it with chrysanthemums, coneflowers, monardas, blazing stars, asters, boltonia, and Indian grass.

Cultivars 'The Blues' was selected for its good blue-green foliage color. Zone 3.

Schizachyrium scoparium var. scoparium

Sorghastrum nutans
Indian grass
Zone 4

Native Habitat Moist to dry prairie and open woodland in the southwestern half of Minnesota.

Height 5 to 7 feet

Description Indian grass is an upright grass with flat, narrow leaf blades and terminal bronzy yellow flowers in late summer and early fall. The flowers have showy bright yellow anthers. Fall color is golden yellow to dark orange, in nice contrast to the golden brown silky tassels of seeds.

Landscape Use The silky soft, golden seed heads of Indian grass impart a special beauty and drama to prairie gardens in autumn and it makes a powerful late-season statement in mixed borders. Finches and sparrows feast on seeds all winter.

Site Prefers a slightly moist to well-drained soil in full sun but will tolerate very light shade.

Culture Space plants 1 to 2 feet apart. Mow or burn this warm-season grass in late winter. Plants tolerate drought once mature. It is not overly aggressive.

Good Companions Indian grass should be planted with little bluestem and other late-summer and fall prairie plants such as asters, boltonia, heliopsis, rudbeckia, wild bergamot, and Joe-pye weeds.

Cultivars 'Sioux Blue' is an upright form with metallic blue summer foliage. Zone 4.

Sorghastrum nutans

Spartina pectinata
Prairie cordgrass
Zone 4

Native Habitat Wet prairies and swamps throughout most of Minnesota.

Height 6 to 9 feet

Description This warm-season grass has a graceful, arching form. The seed heads start out purple in summer then turn yellow gold in fall. Fall foliage color is yellow.

Landscape Use Prairie cordgrass is good in naturalistic plantings and for stabilizing stream banks and pond edges, where it thrives in the wet soil.

Site Requires a moist, fertile soil in full sun. Even moisture is necessary, and it can even be submerged for short periods.

Culture For stream-bank erosion control, space plants 2 feet apart and apply mulch or erosion fabric to hold soil during establishment. In about two years, the entire area will be a solid mat of rhizomes that defy erosion. Prairie cordgrass is too invasive for most landscape situations, but it will spread slower in a heavy soil.

Good Companions Plant prairie cordgrass in moist soils along with other tall moisture-lovers such as Joe-pye weeds and autumn sneezeweed.

Cultivars 'Aureomarginata' ['Variegata'] has long, graceful, ribbon-like foliage striped with yellow bands and good fall color. Zone 4.

Spartina pectinata 'Aureomarginata'

Sporobolus heterolepis
Prairie dropseed
Zone 3

Native Habitat Mesic to dry prairies and savannas in the southwestern half of Minnesota.

Height 2 to 4 feet

Description Prairie dropseed is a clump-forming grass, slowly expanding to form a fountainlike mound about 18 inches in diameter. It has narrow individual blades that are bright green in summer, turning yellow and orange in fall. It blooms late summer, producing many upright flower stalks topped with pale pink panicles that have a luscious scent reminiscent of fresh popcorn and later turn gold.

Landscape Use This graceful, well-behaved grass will add a touch of elegance to any planting. Use it in perennial gardens, mixed borders, or foundation plantings. Space plants 18 to 24 inches apart for distinctive borders. A mass planting is a beautiful site in late summer. The seeds are an important food source for birds in fall and winter.

Site Prefers a well-drained soil with moderate moisture levels in full sun but will tolerate drier conditions. Avoid constantly wet soils.

Culture Slow to grow from seeds; start with plants if possible. Cut back this warm-season grass in early spring before new foliage emerges. Dig and divide clumps in spring as needed.

Good Companions Plant fine-textured prairie dropseed with butterfly weed, coreopsis, asters, coneflowers, and blazing stars in perennial borders or prairie gardens.

Other Species *S. asper* var. *asper* (rough dropseed) and *S. cryptandrus* (sand dropseed) are two similar species sometimes included in prairie seed mixtures. Zone 3.

Sporobolus heterolepis

Additional Native Grasses

Buchloe dactyloides (buffalo grass) is native to dry prairies in the far southwestern corner of Minnesota. It is a warm-season, sod-forming grass growing about 6 inches tall with grayish green foliage. It tolerates heavy soils and is drought tolerant. It can take regular mowing and is sometimes used as an alternative to non-native lawn grasses. Cultivars developed for lawns in southern states are usually female plants and do not have showy flowers. Buffalo grass is rare in Minnesota and is on the state's special concern list. Zone 4.

Calamagrostis canadensis (blue joint) is native to marshes, bogs, and lakeshores throughout Minnesota. This cool-season, rhizomatous grass grows 2 to 5 feet tall and often makes a solid stand that excludes other plants. It blooms June to July with highly variable flowers. Use it over septic tanks, in bog gardens, and along streams and large ponds. It requires moist or wet soil in full sun and can even be partially submerged for part of the year. Zone 3.

Glyceria striata (fowl manna grass) is native to forests, swamps, and lake edges throughout most of Minnesota. This delicate, 2- to 3-foot, clump-forming grass has green flowers July to August and nicely textured foliage. Grow it in moist to wet soil in part sun to full shade. It is excellent for habitat restoration and soil stabilization in low areas and along pond edges and stream banks, where it provides food and shelter for waterfowl. It is also a good candidate for a large bog garden. Mix the small seeds with an inert carrier such as sawdust or vermiculite to ensure good coverage when sowing. Plants will reseed. Zone 3.

G. canadensis (rattlesnake grass) is native to shallow water along lakes and ponds in the northeastern section of Minnesota. It is named for the flower clusters, which resemble rattlesnake tails. Zone 3.

G. grandis (tall manna grass) is native to shallow water along lake margins throughout most of Minnesota. It grows 4 to 6 feet tall with attractive, open, purple flower clusters. Zone 3.

Hierochloe odorata (sweet grass) is native to wet meadows, low prairies, marsh edges, bogs, stream banks, and lakeshores throughout Minnesota. The glossy leaves have a pleasant vanilla fragrance. The golden flowers turn to seed by July but leaves remain green until winter. Growing 1 to 2 feet tall, it prefers a moist soil in full sun near a pond or stream bank. It slowly creeps by rhizomes to form a patch; encircle plants with a root barrier to restrict growth. Zone 4.

Koeleria macrantha (June grass) is native to prairies and savannas in all but the northeastern corner of Minnesota. It is a cool-season bunch grass that blooms earlier than most prairie grasses, usually in June. It grows 1 to 2 feet tall with yellowish or greenish white flowers that are quite showy. Use it in prairie gardens and restorations. Zone 3.

Oryzopsis hymenoides (Indian rice grass) is native to dry soils in northwestern Minnesota. It is a tufted grass growing 1 to 2 feet tall and wide with twisted, wiry stems that bear white, hairy flowers in spring. Grow it in full sun to light shade in well-drained, sandy soils in prairie gardens, rock gardens, and naturalized sites. Zone 3.

Stipa spartea (porcupine grass) is native to prairies and savannas mainly in the southwestern half of Minnesota. It grows 2 to 4 feet tall with straw-colored leaves and silvery, sharp-pointed seeds in summer. Plant it in average to dry soil in prairie gardens or restorations. Zone 3.

Special-Use Native Grasses and Sedges

There are many other native species suitable for prairie restorations and other specialized uses such as wetland sites. These are usually available as seeds, often as part of mixtures, from specialty nurseries. This list includes:

Agropyron trachycaulum (slender wheat grass)
Ammophila breviligulata (beach grass)
Beckmannia syzigachne (American slough grass)
Bromus ciliatus (fringed brome)
Bromus kalmii (prairie brome)
Carex species (sedges)
Cinna arundinacea (wood reed grass)

Cyperus schweinitzii (rough sand sedge)
Danthonia spicata (poverty oat grass)
Diarrhena americana (beak grass)
Eleocharis acicularis (spike rush)
E. palustris (great spike rush)
Eragrostis trichodes (sand lovegrass)
Juncus species (common rushes)
Leersia oryzoides (rice cut grass)
Melica nitens (tall melic grass)

Muhlenbergia species (satin grasses)
Paspalum ciliatifolium (hairy lens grass)
Poa palustris (fowl bluegrass)
Scirpus species (hard-stem bulrushes)
Scleria verticillata (low nut rush)
Sparganium eurycarpum (giant bur-reed)
Stipa viridula (green needle grass)

Ferns

Ferns evoke the essence of Minnesota's northwoods, where most native ferns grow in moist, rich, acidic soil in part to full shade. They are actually a conspicuous part of the state's vegetation in all but the driest prairie ecosystems. Many adapt well to shady landscape situations, where their foliage provides interesting textural contrasts to other plants. They add visual buoyancy, lightening the garden with their wide assortment of foliage shapes in varying shades of green. With adequate soil and moisture, fern foliage will be attractive until fall frost kills them back or winter snow covers them.

Ferns are typically used in shade gardens, where they are excellent background plantings, fillers, blenders, groundcovers, or focal points. Although the small fiddleheads appear fairly early in spring, the leaves do not fully mature for quite awhile. This makes ferns good companions for many of the early flowering ephemerals and spring bulbs, which die back in early summer, leaving room for the ferns to fill in. Some can be used in foundation plantings, and some of the sun-tolerant types can be used in mixed borders. Ferns are fun for collectors, and some gardeners choose to devote gardens solely to these plants. Ferns look good around water gardens, and some of the smaller types can be used in shady rock gardens. They can be used to create a soft boundary between one section of a garden and another. They can even be used to form a small hedge in the right spot.

The most recognizable feature of a fern is its frond, or leaflike structure. Fronds range from $1/16$ of an inch up to several feet long and may be feathery to coarse in texture. They have two main parts: the stipe, or leaf stalk, and the blade, which is the leafy portion. Fronds are often divided and subdivided and sometimes divided again. Most ferns have two different shapes of fronds—sterile and fertile—and these can be quite different on some species.

Most ferns grow from underground perennial rhizomes, which can lead to large stands of a single species. They push up new fronds in tightly curled fronds called fiddleheads. This method of growth protects fern tips during emergence. Instead of growing from seed like most flowering plants, ferns come from a single spore that develops into the sporophyte. Spores are borne in a spore case. The case contains many individual spores and is usually found on the underside of a frond or on separate stalks. Inexperienced gardeners often become concerned over these fruiting bodies and assume their plants are infested with a rust disease or unusual insect.

Planting

Native ferns are easy to grow if a few simple cultural requirements are met. Most prefer filtered shade as opposed to deep, heavy shade. Dappled shade, where patterns of sunlight move across the plants as the day progresses, is best. In some cases you may need to limb up some trees to allow more light to reach ferns. Some ferns will grow in full sun if sufficient moisture is provided. In general, the larger and more robust the fern, the more it will tolerate sunlight. Most need protection from midday sun.

Most ferns grow best in slightly acidic, humus-rich soil. Dig or till at least 6 to 8 inches deep and add 3 to 4 inches of organic matter in the form of peat moss, ground pine bark, or shredded leaves. Well-rotted manure and compost also work.

When planting, do not allow roots to become dry. Bare-root ferns should be unpacked immediately upon arrival and their roots wrapped in wet newspaper and placed in the shade out of any wind until they can be planted, which should be as soon as possible. Plant bare-root ferns in fall or early spring only. Plant as any perennial; dig a hole, spread out rhizomes, and plant the crown no deeper than 1 inch below the soil surface. Remove any broken fronds, as they will not be repaired. Container-grown ferns can be planted at any time during the growing season if you can provide the young plants with adequate water and protection from sun. Plant at the same depth as in the container. Space small ferns about 1 foot apart, medium-sized ferns about 2 feet apart, and large ferns 3 feet apart.

Care

Ferns should never be allowed to become bone dry. Plant them in an area where you will be able to supply supplemental water during droughts. Water deeply several times a week during dry spells, right up until they go dormant for winter. An organic mulch applied at planting time will help conserve water as well as retard weed growth. Good

Ferns are the backbone of shade garden. Not only do they offer color and texture throughout the summer months when most shade plants are less showy, they also serve to fill in the bare spots when spring ephemerals fade away.

mulches are shredded bark, pine needles, and dry oak leaves. Dig and divide ferns every few years in spring to keep plants vigorous. Replant divisions with the crowns at soil level.

In spring, gently brush away any leaves that have matted over ferns and crumble up by hand any erect dead fronds. Add mulch if needed, and the ferns are ready to go for the season. They shouldn't need additional fertilizer if you allow leaves to remain on the bed each fall. Pull weeds by hand. Don't use a hoe or tiller. Rhizomes are close to soil surface and can be easily injured.

Possible Problems

Most ferns are not bothered by diseases, but slugs can be a problem. Unfortunately, there is no foolproof method for eradicating these slimy pests. All you can really hope to do is reduce their numbers and protect plants when

they're young and most vulnerable. You can install physical barriers, such as plastic-bottle cloches, or sprinkle lime, eggshells, or sawdust around plants. Slugs are attracted to saucers or plastic pots of milk or beer. Lay rolled-up newspaper or boards on the soil. Slugs will seek out these protected areas and can be easily removed. Drop them in a bucket of a 5 percent alcohol-water solution or soapy water, or spray the slugs with an ammonia-water solution. Remember that toads, frogs, and beetles eat slugs and are worth encouraging in your garden.

Several fern species are in jeopardy in Minnesota and must not be dug from the wild. As with all native plants, purchase native ferns only from nurseries selling nursery-propagated plants. Growing nursery-propagated plants in your garden will help to maintain biodiversity and help ensure the survival of endangered species.

Adiantum pedatum
Maidenhair fern
Zone 3

Adiantum pedatum

Native Habitat Moist, well-drained woodlands in eastern Minnesota.

Height 12 to 24 inches

Description Maidenhair fern has horizontal, lacy, fan-shaped, arching branches and wiry, black stems. The individual fronds are green and fan-shaped with black petioles and they turn golden yellow in fall.

Landscape Use This delicate fern is attractive spring through fall. Its fine texture adds softness to the landscape. It can be grown in the dry shade under large trees or in shady rock gardens.

Site Prefers moist, slightly acidic, rich soil in partial shade, but tolerates all but very dry soils.

Culture Maidenhair fern is easy to grow in gardens. Transplant it carefully to avoid damaging the thin stems. Plant it in groups spaced about 18 inches apart or tucked here and there. With time, 2-foot-wide clumps will cover the ground.

Good Companions Fine-textured maidenhair fern combines well with almost all woodland plants, but is especially effective with coarse-textured plants such as hostas.

Athyrium filix-femina var. *angustum*
Lady fern
Zone 2

Athyrium filix-femina

Native Habitat Wide range of habitats throughout most of Minnesota.

Height 2 to 3 feet

Description Lady fern forms a cool green carpet of lacy, deeply cut fronds that arch from the crown. The bright green to light yellow, 5- to 12-inch fronds are produced continually during growing season, keeping plants fresh-looking.

Landscape Use Lady fern is an excellent groundcover in moist woodland gardens, where it makes a nice backdrop for smaller plants. It can be naturalized along stream banks or shady water gardens. It is large enough to be used with shrubs in foundation plantings.

Site Prefers a slightly acidic, moist to wet soil in partial to full shade, but tolerates a wide variety of soils. Can take some sun if the soil is kept moist.

Culture Lady fern is easy to grow provided plants receive consistent moisture. Mulch plants and water during dry periods. The fronds are somewhat brittle, and older fronds may become tattered by late summer if they are grown in high-traffic areas or windy sites. It is slow spreading.

Good Companions Use lady fern to brighten a shady spot. Its light color sets it off nicely from other ferns, and it is a good backdrop for bloodroot, trilliums, and other native flowers.

Cultivars 'Frizelliae' has narrow fronds (less than 1 inch wide) with fuzzy, green, earlike projections along the sides. It is an interesting specimen plant for a special spot in woodland gardens. Zone 3. 'Vernoniae Cristatum' and 'Victoriae' were selected for the attractive crests at the ends of their leaves. Zone 4.

Cystopteris bulbifera
Bulblet fern
Zone 3

Native Habitat Moist, rocky woods and along streams in the eastern half of Minnesota.

Height 12 to 30 inches

Description Bulblet fern has narrow, lacy, twice- or thrice-cut triangular fronds that are widest at the base. Bulblets form along its main stem and fall off to start new plants. The young stems are dark red when they emerge in spring, offering interesting color contrast to other spring plants.

Landscape Use Grow bulblet fern in a shady nook in rock gardens or in a moist spot in shade gardens. In its native habitat it is often found clinging to rocks wet with water spray, so it would look at home near pools or waterfalls. It grows in humus-packed limestone crevices.

Site Requires a moist, rich soil with a pH of 7.0 to 7.5 in shade.

Culture Bulblet fern is easy to establish, and eventually forms large colonies. Young plants are easy to cull out, however.

Good Companions Place bulblet fern near water features with other moisture-loving plants such as cardinal flower, obedient plant, and blue flags; good fern companions include *Dryopteris* species and *Polypodium virginianum*.

Other Species *C. protrusa* (protruding fragile fern) is native to moist, shady slopes in the far southeastern corner of Minnesota. It grows 10 to 12 inches tall in clumps 2 feet or more across. The fronds are 10 inches long and 3 inches wide. It continually forms new fronds, and the fresh, light green clumps attract attention in shade gardens, especially in spring. It grows best in a slightly acidic to neutral soil in moist shade. Extra water during the growing season will keep plants green and encourage production of new fronds. Zone 4.

Cystopteris bulbifera

Dryopteris marginalis
Marginal shield fern
Zone 4

Native Habitat Rich, rocky woods in the far southeastern corner of Minnesota.

Height 18 to 24 inches

Description Marginal shield fern is a large, leathery woodland fern with fronds borne in a crownlike cluster 18 to 24 inches long and 6 to 10 inches wide. The fiddleheads are densely covered with golden brown hairs.

Landscape Use Use marginal shield fern in shady rock gardens or woodland gardens. The evergreen foliage provides year-round interest.

Site Prefers a cool, shady site protected from sun and drying winds.

Culture Incorporate lots of organic matter into the soil before planting. Provide water during dry periods. Marginal shield fern reaches the northern and western limits of its range in Minnesota, so it may struggle during cold winters. It is on Minnesota's threatened species list, so only purchase nursery-propagated plants.

Good Companions The dark green fronds of marginal shield fern set off white and red flowers nicely. Grow it with cardinal flower and trilliums. Other good companions include wild sarsaparilla, large-leaved aster, bluebead lily, and Canada mayflower. In fern gardens, place it near bladder ferns (*Cystopteris* spp.), wood ferns (*Dryopteris* spp.), and common polypody (*Polypodium virginiana*).

Other Species *D. carthusiana* (spinulose shield fern) is native to damp wooded areas throughout most of Minnesota. It is an erect fern with outward-curving evergreen fronds that are often used in floral arrangements. Give it partial shade and moist, acidic soil and it will perform nicely as an accent plant or a mass planting in woodland gardens. Zone 3.

D. cristata (crested fern) grows in moist woodlands northeast of a diagonal line running across Minnesota. The erect, 18- to 30-inch fertile fronds have horizontally arranged, widely spaced pinnae and a slight twist to them. Grow it in light shade in moist soil. Zone 3.

D. goldiana (Goldie's fern) is rare

Dryopteris marginalis

in Minnesota, found in cool, moist woods and shady ravines widely scattered in the southeastern corner of the state. It is a large, impressive fern growing 3 to 5 feet tall in clumps up to 2 feet across and makes a spectacular clump in woodland gardens. It requires a cool, moist, shady location in slightly acidic, humus-rich soil. It is on Minnesota's special concern list. Zone 4.

Matteuccia struthiopteris var. *pensylvanica*
Ostrich fern
Zone 4

Native Habitat Swamps and wet woods throughout Minnesota.

Height 2 to 5 feet

Description Ostrich fern is vase-shaped with large, plumelike, leathery sterile fronds and smaller fertile fronds that appear in late summer. The fertile fronds become brown and woody in fall and persist through winter, offering interest even into early spring.

Landscape Use Ostrich fern has a regal look to it and can be used in more-formal landscape situations such as north-facing foundation plantings, shade gardens, and mixed borders. The young fiddleheads are up early in spring. They are edible, with a taste similar to asparagus.

Site Prefers a cool, wet, slightly acidic soil in part to full sun but is adaptable to drier conditions.

Culture Ostrich fern is easy to grow if given suitable conditions. It spreads readily by rhizomes, but it is fairly easy to keep in check. Leaves will scorch if soil becomes dry.

Good Companions Ostrich fern can be planted with shade-tolerant shrubs and it is a good choice for filling in after spring bulbs and ephemerals are gone.

Matteuccia struthiopteris var. *pensylvanica*

Onoclea sensibilis
Sensitive fern
Zone 2

Native Habitat Wet, open woods, swamps, and other moist areas in northern and eastern Minnesota.

Height 12 to 30 inches.

Description Sensitive fern has interesting deeply pinnate fronds to 18 inches. They are light green, large, and lobed, somewhat resembling a green glove. The sterile fronds die with the first frost, but fertile fronds persist through winter. Fiddleheads emerging from rhizomes form a distinctive pale red mass in spring.

Landscape Use Sensitive fern is best for large gardens as a mass planting. It will grow in still water along ponds and stream banks, even in full sun.

Site Prefers wet to slightly moist, acidic soil in light shade; sun is okay if the soil is quite wet.

Onoclea sensibilis

Culture Sensitive fern produces long, robust rhizomes and may become invasive in the right conditions.

Good Companions Sensitive fern is best planted in masses or with other large ferns as groundcover.

Osmunda cinnamomea
Cinnamon fern
Zone 3

Native Habitat Wet soil of wooded swamps, marshy places, and wet savannas generally throughout the eastern half of Minnesota.

Height 30 to 36 inches

Description This upright fern has waxy, deep green foliage. The fertile fronds occur in spring just above the foliage as a cinnamon-colored spike in the center of the clump of hairy fiddleheads, which can reach up to 8 inches before unfolding. The fertile fronds are soon replaced by sterile fronds that are yellow green and deeply pinnate with a dense tuft of rusty hairs beneath the base of each pinna.

Landscape Use This large fern makes a bold statement in the landscape. Use it as a background planting or along water's edge. It can also be used in foundation plantings or as a background for flowers. It is difficult to see individual plants in established plantings.

Site Does best in acidic, moist soil and partial shade but adapts fairly well to a wide range of conditions, including full sun if the soil is continually wet.

Culture Easy to grow. Requires little care once established.

Good Companions Plant in masses or with other large ferns in fern gardens. It can be used as a backdrop for cardinal flower and turtleheads in moist-soil areas.

Other Species *O. claytoniana* (interrupted fern) is native to rich, shaded woods generally in the eastern half of Minnesota. The upright-growing, vase-shaped deciduous clump looks a bit like cinnamon fern, except that the normal fronds along the stem are suddenly "interrupted" by a few brown fertile fronds. Culture and use are the same. Zone 2.

O. regalis var. *spectabilis* (royal fern) is native to swamps and moist, acidic soils in northeastern Minnesota. It grows 24 to 36 inches tall or more and is coarser than the other two species. The fronds unfurl with a wine-red color before turning green. The light green, leathery leaves give it a tropical appearance, and it looks especially nice planted near water, where it thrives in the damp soil. Zone 3.

Osmunda cinnamomea

Osmunda claytoniana

Osmunda regalis

Polystichum acrostichoides
Christmas fern
Zone 3

Polystichum acrostichoides

Native Habitat Moist to well-drained upland soils in the far southeastern corner of Minnesota.

Height 18 to 24 inches

Description Christmas fern has dark green evergreen fronds. The fertile fiddleheads are covered in silvery scales. Smaller sterile fronds are produced later in the season.

Landscape Use Christmas fern can be massed as groundcover, or a clump looks nice tucked into the base of a stump. The evergreen foliage adds year-round interest but looks rather tired by spring. It is often used for cut greenery during the holidays.

Site Prefers fertile, well-drained woodland soil in partial to almost full shade.

Culture Plants are easy to establish if grown in partial shade in constantly wet soil. This clump-former stays nicely in place. It is on Minnesota's threatened species list, so only purchase nursery-propagated plants.

Good Companions Christmas fern's dark green color is a good backdrop for smaller woodland plants.

Other Species *P. braunii* (Braun's holly fern) is native to cool, shaded gorges along the Katunce River in far northeastern Minnesota, where it grows on small ledges and cracks above the spring high waterline. It is rare and is on Minnesota's endangered species list. The bipinnate fronds taper to the base and tip. Each frond has twenty to forty pairs of pinnae, which are sharply serrated. It requires cool woodland soil that has good moisture retention. Zone 3.

Additional Native Ferns

Asplenium platyneuron (ebony spleenwort) is native to moist woods, often on rocks, in the far southeastern corner of Minnesota. It grows 8 to 18 inches with spreading sterile fronds that lie close to the ground and showy erect fertile fronds. It is an excellent smaller fern for shaded rock gardens or planted among rocks bordering woodland paths. Grow it in a loose, woodsy, near-neutral pH soil in partial shade. Once established, ebony spleenwort volunteers frequently and generally maintains itself with little additional care required. It will tolerate dry periods. It is on Minnesota's special concern list, so only buy nursery-propagated plants. Zone 4.

A. rhizophyllum [Camptosorous rhizophyllus] (walking fern) is native to damp, shady places in the far southeastern corner of Minnesota. It is named because the tips of the long-tapering, 4- to 12-inch fronds arch over and root, producing new plantlets as it "walks" along the ground. Zone 4.

A. trichomanes var. trichomanes (maidenhair spleenwort) is found in ledges and crevices in moist, east-facing cliffs mainly in the far northeastern corner of Minnesota. It is a small, delicate fern growing 3 to 6 inches tall with spreading leaves, a good choice for cool, moist rock gardens. It is difficult to find sources for this plant, and it is on Minnesota's threatened species list, so only buy nursery-propagated plants. Zone 3.

Cheilanthes lanosa (hairy lip fern) is found only in far east-central Minnesota. Eventually reaching only 8 inches in height with a spread of 20 inches, this extremely drought-tolerant fern is composed of durable olive-green fronds. It grows equally well in a sunny spot or in light shade and can tolerate acid or alkaline soils and brief dry periods once established. It is on Minnesota's endangered species list. Zone 4.

Pellaea atropurpurea (purple cliff brake) grows in limestone cliffs in the far southeastern corner of Minnesota along the Mississippi River. It has gray to bluish green, 6- to 20-inch, pinnately compound fronds. Plant it in pockets of humusy soil in limestone in light shade to partial sun in crevices in large boulders or in rock walls or ledges. It is on Minnesota's special concern list, so only buy nursery-propagated specimens. Zone 4.

Polypodium virginianum (rock-cap fern, common polypody) is native to rocks and cliffs mainly in the eastern half of Minnesota but also farther west. The rhizomes grow in a mat partially exposed and the leaves form a dense, yellow-green mass. The deeply pinnate, 4- to 12-inch fronds are leathery and evergreen. Use rock-cap fern to soften ledges and crevices of rock walls or to cover large boulders in woodland gardens. It can also be grown in the rock garden, where it grows well in crevices between rocks in open shade. It can take several years for rock cap fern to permanently establish itself. Mats can be placed on suitable rock surfaces with a bit of humus and weighted down. Supply additional moisture in periods of drought to keep plants in good condition. Zone 3.

Pteridium aquilinum var. latiusulum (bracken) is native on well-drained soils of thin woods in all but the southwestern corner of Minnesota. It grows 12 to 48 inches tall with coarsely textured, broadly triangular fronds. It tolerates well-drained soil in partial sun and can be used as a groundcover along roadsides or at the edge of a woodland. It may become invasive under favorable conditions, but is easily controlled by mowing or cultivation. Zone 3.

Thelypteris palustris var. pubescens (northern marsh fern) is native to marshes and bogs in all but the southwestern corner of Minnesota. It grows 18 inches tall and has light green, lacy fronds, each arising individually from a creeping rhizome without forming clumps. It spreads rapidly, especially when given supplemental water or when grown in damp soil. Use it to edge streams and ponds. Zone 3.

Woodsia obtusa (blunt-lobed woodsia) is native to rock crevices in southern Minnesota. It has deep green fronds and grows 6 to 18 inches tall. Use this slow spreader in rock gardens or wall gardens. Plants prefer a neutral soil rich in organic matter. Once established, they will survive prolonged dry periods. Zone 3.

Pteridium aquilinum

Deciduous Trees

Planting a tree is a long-term investment, but one that returns a great deal of satisfaction and value. Since most trees are not easily moved once they are established, you should make sure the tree you have selected is a good match for your site. Especially keep in mind the mature size. Don't make the mistake of planting a tiny oak seedling 10 feet from your front door or a basswood under a powerline. Also consider hardiness, light requirements, rate of growth, and seasonal aspects like flowers, fruits, fall color, and winter appearance. You will probably want to select a tree based on your soil conditions if you don't want to drastically change the soil. Consider if you want to attract wildlife and birds. Some trees can also be messy, dropping leaves, flowers, fruits, or twigs.

Quercus rubra (red oak) is one of the best native trees for landscape use, offering strong wood and good-colored foliage both in the growing season and in winter.

Planting

You will have the best success planting northern-grown nursery trees that have been properly root pruned. They will survive transplanting the best and start growing quickly. If you have the opportunity to move a native tree, stick with one 1 or 2 inches in diameter or less and move it in early spring. Bare-root trees must be planted in spring as soon as the soil can be worked. Balled-and-burlapped, container-grown, and tree-spade trees can be planted anytime but the hottest days of summer, during July and August. Spring is still the best time for planting, however.

Before planting, amend the soil with a good amount of organic matter such as compost, peat moss, or well-rotted manure. Mix this thoroughly with the planting-hole soil. Place the tree at the same depth it was growing at in the container or burlap wrap. Bare-root trees should be planted so that the crown is level with the ground level. Newly planted trees should not need additional fertilizer. Do not stake them unless they are on an extremely windy, open site. Any stakes should be removed as soon as the tree has rooted well, usually after the first year. It is a good idea to surround all newly planted trees with a ring of organic mulch 2 to 4 inches thick. Good mulches are wood chips, shredded bark, and pine needles. Replenish the mulch as needed throughout the growing season. Do not use rock or black plastic as mulch.

Care

The first two or three years after planting, make sure the soil is evenly moist from spring until the time the ground freezes in fall. Once established, most trees can tolerate some dry periods, but don't hesitate to water as needed, especially in sandy soils. Always saturate the soil thoroughly with each watering to encourage deep rooting. Most young trees will also benefit from a spring application of fertilizer. Spread a layer of rotted manure or compost around each tree or use a fertilizer such as Milorganite or fish emulsion. If possible, allow leaves to fall and decay under trees to return nutrients to the soil. Keep weeds pulled or smother them with organic mulch.

Maintenance needs of trees differ with each species. Most deciduous trees will benefit from some selective pruning while they are young to develop a good shape. The best time to do this is usually while they are dormant: you can see their silhouette better, and it reduces the chances of disease and insect problems. Remove branches that have narrow-angled crotches as these branches are weaker than wide branches and are likely to split as the tree gets older. Remove lower branches only if they interfere with the use of the areas under the tree. Remove any dead, damaged, or diseased parts of trees at any time of year, cutting back to just above a healthy, outfacing bud.

Possible Problems

Some thin-barked trees such as maples and mountain ashes need to be protected from winter sunscald while they are young. Trees are susceptible to sunscald until they have developed a thick, corky trunk. Wrapping trees in fall with a tree wrap made of weather-resistant paper minimizes sunscald. Young trees should also be protected from rodent feeding, which can kill a tree. The best way to do this is to surround the trunk below the first branch with hardware cloth.

Some native trees are susceptible to insects and diseases, some of which can be fatal, such as oak wilt and Dutch elm disease. However, most insect and disease problems are purely cosmetic and don't really threaten the tree's life. On large trees, control measures are rarely practical. Some things to watch for on young trees are aphid feeding (spray leaves daily with a strong blast of the hose), iron chlorosis (acidify soil to lower pH), and leaf scorch (water trees well during dry periods and mulch soil). The best defense against insect and disease problems is a vigorously growing tree. By choosing a tree appropriate for your site conditions and providing it with the necessary nutrients and ample water, most pest problems will not become serious.

Best Native Deciduous Trees for Landscape Use

Acer rubrum (red maple)	*Ostrya virginiana* (ironwood)
Acer saccharum (sugar maple)	*Quercus alba* (white oak)
Betula nigra (river birch)	*Quercus rubra* (red oak)
Celtis occidentalis (hackberry)	*Tilia americana* (basswood)
Gymnocladus dioica (Kentucky coffee tree)	

Acer rubrum
Red maple
Zone 4; most of zone 3

Native Habitat Wet to moist soils along swamps or depressions in northeastern and east-central Minnesota.
Size 40 to 65 feet tall, 45 feet wide
Description Red maple is a medium-sized tree with a broadly rounded symmetrical crown. The smooth, light gray bark on young stems turns dark gray and shaggy on older limbs. Leaves have three- to five-pointed, saw-toothed lobes. The upper surface is light green, the lower surface whitish and partly covered with pale down. It is the first of the native maples to turn color in fall. Fall color is usually a brilliant red, but it can also be orange or yellow. The small red flowers give the bare branches a red glow for a week or so in early to mid-April, before the leaves appear.
Landscape Use Red maple is a good shade, lawn, or street tree. Its ability to survive in heavy soils means it can tolerate the poorer, compacted soils of streets and parking areas. The red flowers are a welcome site in early April when few other trees are showy. All maples have shallow roots and produce deep shade that can make it difficult to grow grass under.
Site Prefers slightly acidic, sandy loam soils that are well drained in sun to light shade, but it is tolerant of most landscape conditions. It will not grow well on alkaline soils, but it will tolerate moderately moist soils.
Culture This tree is moderately fast growing and tolerant of urban conditions. It has a wide native range; stick with northern seed strains and sources for best results. Wrap young trees in winter to protect them from sunscald. Fall color can be inconsistent; purchase trees in fall if you want good fall color. Avoid pruning trees in late winter when the sap has begun to flow. This "bleeding" sap does not harm the tree, but it is messy and unsightly.

Acer rubrum

Cultivars 'Autumn Spire' was selected from a population near Grand Rapids, Minnesota, for its bright red fall color and columnar form. Zone 3. 'Northwood' is a University of Minnesota release selected from a native population near Floodwood, Minnesota, that is fast growing, nicely shaped, and has good fall color. Zone 3.

Other Species *A. nigrum* (black maple) is native to moist, fertile soils, floodplains, and bottomlands in the southeastern quarter of Minnesota. It is similar to *A. saccharum* but the leaves droop at their sides and have yellow fall color. It has greater heat and drought tolerance than sugar maple and is a better choice for the western part of Minnesota. Zone 4.

A. saccharinum (silver maple) is native to wet to moist soils in the eastern half of Minnesota. It is a fast-growing species with weak wood, which makes it a messy tree after windstorms. It is tolerant of a wide variety of landscape situations and is often used in western Minnesota where other maples don't do as well. The leaves are deeper cut than other maples and are whitish on the undersides. Fall color is usually pale yellow. It grows 80 feet tall or more, so keep it away from small yards and buildings. Older bark peels in vertical strips. It produces many seeds, which can become weedy, and the shallow, invasive root system makes it difficult to garden under. It is best used on tough, open sites where quick shade is needed. Zone 3.

A. saccharum (sugar maple) is native to rich, moist soils in the eastern half of Minnesota. It is an excellent choice for landscape use where the soil conditions are right. It prefers heavy clay or loam soils that are moisture retentive and on a north-facing slope, but it will tolerate drier, sandier sites. Avoid compacted, alkaline soils. It is larger than red maple, growing 50 to 80 feet with a nice round canopy. Its greatest attribute is the brilliant fall color, which ranges from yellow to orange to scarlet. It is slow growing when first transplanted. It is sensitive to salt damage, so avoid using it as a street tree. It is also susceptible to leaf scorching and tattering when planted on open, exposed sites, which can make the leaves somewhat unattractive in late summer. Zone 4; protected sites in zone 3. 'Legacy' was selected for its good fall color and resistance to leaf tatter. Zone 4.

A. spicatum (mountain maple) is native to southeastern and northeastern Minnesota on moist soils along streams and other wet areas. It is low branched and only grows to about 25 feet tall and 20 feet wide, making it more shrublike, but it can be pruned to a small tree. It has a crooked trunk and upright branches. Fall color is a beautiful deep orange to red. It prefers moist, acidic soils in partial shade. Being an understory tree, it doesn't do well on open, exposed sites. If you can provide it with the right conditions, it is a good substitute for the overplanted nonnative amur maple (*Acer ginnala*). Zone 3.

Acer saccharum

Acer spicatum

Betula nigra
River birch
Zone 4; trial in zone 3

Native Habitat Rich lowlands of streams and rivers in far southeastern Minnesota.

Size 40 to 60 feet tall, 15 to 25 feet wide

Description River birch is an upright tree with a rounded or irregularly spreading crown. It can be grown as a single- or multi-trunked tree. The distinctive bark is highly ornamental; it is shiny and varies in color from red brown to cinnamon brown with a touch of salmon pink. It peels off in large horizontal strips. On older trees, the bark is darker and corkier. The toothed leaves are roughly triangular, with a dark green upper surface and pale yellow green underneath. Fall color is yellow to green and is not as ornamental as other birches.

Landscape Use River birch offers landscape interest year round thanks to its unique peeling bark. It is one of the few large trees that can be grown as a clump. Use it as a shade or specimen tree.

Site Prefers moist soils and tolerates wet soils, but will do well on upland sites if given supplemental water when young. Needs full sun and a soil pH below 6.5 to prevent chlorosis. It is the most heat tolerant of the birches, making it the best choice for most landscape situations.

Culture Select plants from a northern seed source. Acidify soil before planting if needed. Maintain adequate soil moisture by surrounding plants with a 2- to 4-inch layer of organic mulch. Prune in summer to avoid bleeding. The brittle, twiggy branches break easily in windstorms, and catkins dropping in late spring can be messy, but only for a short period. River birch is highly resistant to bronze birch borer and is rarely troubled by leaf miners.

Cultivars Heritage® ('Cully') is a popular cultivar (perhaps too popular) with lighter-colored bark and larger, glossier, dark green leaves. It usually grows to only about 45 feet tall. Fox Valley® ('Little King') is an interesting dwarf selection that only grows 8 to 10 feet tall in ten years. Zone 4.

Other Species *B. alleghaniensis* (yellow birch) is native to rich, moist soils in eastern and northern Minnesota. It grows 60 to 80 feet tall so is best used in large landscapes. Bark is yellowish or bronze changing to reddish brown and peels off in thin strips, giving the single thick trunk a shaggy appearance. Autumn color is a beautiful golden yellow. It will grow best on moist soils that are not compacted in sun or light shade. It is somewhat resistant to bronze birch borer. Zone 3.

B. papyrifera (paper birch) is native to moist, acidic soils in all but the southwestern corner of Minnesota. It grows 40 to 70 feet tall with a spread of 20 to 30 feet. It has attractive, bright white bark that peels away in sheets to reveal salmon-colored under-bark. Fall color is a good yellow. It is fast growing and rather short-lived, especially in the landscape. It likes full sun but requires adequate moisture and cool roots, a situation often difficult to find in the landscape. It is susceptible to cankers, wood rots, leaf miner, birch skeletonizer, and bronze birch borer. To reduce problems with bronze birch borer, which can be devastating, plant it on cool northern sites where the roots are shaded. Spread a wide circle of organic mulch such as wood chips or shredded bark under the canopy and make sure it receives adequate water during dry periods. Avoid planting it in a sunny lawn or near hot pavement. Any pruning should be done in late summer to avoid attack by bronze birch borers. Zone 2.

Betula nigra

Betula nigra 'Little King'

Betula papyrifera

Carpinus caroliniana

Carpinus caroliniana fruits

Carpinus caroliniana var. virginianum
Blue beech
Zone 3

Native Habitat Rich, moist woods, especially along streams, mainly in east-central and southeastern Minnesota.

Size 20 to 25 feet tall, 12 to 18 feet wide

Description Blue beech is a bushy small tree or shrub with a spreading irregular crown. The beautiful muscle-like bark is bluish gray, smooth, and sometimes marked with dark brown horizontal bands. Slender brownish catkins dangle from branches in early to mid-April. Leaves are dark green in summer changing to a beautiful orange to red to reddish purple in autumn. The small nutlets hang in clusters and turn brown, adding interest in fall and winter.

Landscape Use Plant blue beech where you can enjoy the attractive blue-gray bark in winter. It is good in naturalized plantings and as an understory tree or shrub in woodland gardens. Use it for screening or as a background planting. It can be sheared into a tall hedge or pruned as a single-stemmed small tree. The early spring catkins are a welcome sight, and the nutlets are food for many birds.

Site Prefers a moist, fertile, slightly acidic soil in part shade or sun. Will tolerate full shade and drier conditions.

Culture Blue beech is somewhat difficult to transplant. Move balled-and-burlapped or container-grown plants in early spring. Fertilize lightly when young and shape with selective pruning to form a single trunk that will showcase the interesting bark. Without pruning, it will send up suckers from the base and become shrubby in appearance. It has no serious insect or disease problems.

Carya ovata var. ovata
Shagbark hickory
Zone 4

Native Habitat Rich, damp areas along streams and on hillsides in the far southeastern corner of Minnesota.

Size 60 to 100 feet tall, 40 to 60 feet wide

Description Shagbark hickory is a large, stately tree with a wide-spreading growth habit. It is named for its showy, shaggy bark that flakes off in thin plates. It has a narrow irregular crown. The deep yellow-green summer foliage changes to rich yellow and golden brown tones in fall and lasts for a long time.

Landscape Use This slow-growing large tree brings a lot of character to the landscape with its peeling, shaggy bark, especially in winter. It is a good specimen tree, but requires a large yard or open area. The tough, strong wood is resilient in wind and snowstorms, making it a good choice for windbreaks and hedgerows. The fruits, which grow up to 1½ inches in diameter, are edible and quite sweet. Squir-rels love them, and if you have a shagbark hickory, you will have an abundance of these furry creatures. Wood chips are used to flavor smoked foods.

Site Prefers well-drained soil in sun; young trees can take some shade. Adaptable to a wide range of soil conditions, including wet sites and acidic soil. The deep taproot allows this tree to survive dry upland sites.

Culture Shagbark hickory is difficult to transplant because of its taproot. Move young plants in early spring or plant hickory nuts in fall where you want them to grow. Protect them with some sort of removable wire cage so rodents don't dig them up before they germinate. The fallen nuts may create a litter problem in the lawn.

Other Species *C. cordiformis* (bitter-nut hickory) is native over a wider range in moist lowlands in southeastern Minnesota. It has an inedible nut that is not attractive to wildlife and the bark is smooth and gray. Fall color is a golden yellow. It isn't nearly as ornamental as shagbark hickory, but it is worth preserving if you have one. Zone 4; trial in zone 3.

Carya ovata

Carya cordiformis

Celtis occidentalis
Hackberry
Zone 3

Native Habitat A wide variety of soils in the southern half of Minnesota and in the western portion on flood plains.
Size 40 to 70 feet tall, 20 to 65 feet wide
Description Hackberry's form is determined by its growing conditions. Depending on how much room it has, it varies from a vase-shaped, upright tree to one with an open, wide-spreading crown. The lance-shaped leaves are bright green and rough on top, lighter green on the bottom. Fall color is yellow to greenish yellow. Best fall color comes when severe weather is delayed. The small fruit turns from orange red to purple in September and often hangs on for much of the winter, providing food for several wildlife species. Mature trees have interesting deep, corky bark with warty protrusions.
Landscape Use Its adaptability to a wide range of conditions makes hackberry a good tree for shade, windbreaks, street use, and shelterbelts. It can tolerate wind, full sun, and the dirt and grime of city conditions, and the rough bark offers protection along city streets. It should be considered as a replacement for the American elm. Hackberry is among the best food and shelter trees for wildlife. Birds and mammals eat the fruits, and leaves are the larval food of many butterflies. The narrow limb crotches and numerous spur branches attract many nesting birds.
Site Prefers moist, well-drained soil but will tolerate both wet and dry sites and a wide range of soil pH. It grows in full sun or partial shade.
Culture Hackberry transplants easily, but sometimes takes up to two years really to start growing. After that, it is moderately fast growing, especially on fertile, moister soils. It is drought tolerant once established. A few cosmetic problems affect hackberry, but none of them are serious. Hackberry nipple gall is a wartlike growth on the lower side of leaves caused by insects known as psyllids. Clusters of twiggy outgrowth called witches' brooms appear on some branches, caused by mite feeding and a powdery mildew fungus, but this condition is not harmful to the tree. Some people actually find these witches'-broom growths interesting.
Cultivars 'Prairie Pride' has thick, leathery, dark green foliage; a nice uniform, compact, oval crown; and it does not develop witches' broom. Zone 3.

Celtis occidentalis

147

Fraxinus pennsylvanica var. pennsylvanica
Green ash
Zone 3

Native Habitat Deciduous forests throughout Minnesota; most abundant in valleys and along streams.

Size 50 to 65 feet tall, 30 to 50 feet wide

Description Green ash is a round-topped tree with spreading branches. The dark brown or gray bark is strongly furrowed. Leaves are bright green or yellowish green. Fall color is inconsistent, usually golden yellow.

Landscape Use This tree's adaptability has made it a widely planted landscape tree, and its overuse has led to many problems. Plant this tree only if there are no other trees suitable for your site and there are not a lot of green ashes in your area. It will grow where many other shade trees won't, offering relatively quick shade. Use it for framing, shelterbelts and windbreaks, shade, and in backyard corner plantings. It tolerates street conditions. Leaves appear late in spring and

drop soon after the first fall frost, so it's a good choice where you want sunlight in spring and fall.

Site Green ash grows best in moist, well-drained soils but also tolerates dry, compacted soils. It is moderately shade tolerant, drought resistant, and alkaline tolerant, making it a good choice for soils with a higher pH.

Culture This large tree is easy to transplant, fairly fast growing, and will withstand severe conditions of both soil and climate. When healthy, it is long-lived. In stressful situations, ashes often develop several disease problems collectively called "ash decline." In the past decade or so, many seemingly healthy ash trees have died rather quickly. This affliction is thought to be a lethal combination of environmental stresses and insect and disease problems that eventually overwhelm the plants. Afflicted trees have premature leaf and flower drop, creat-

ing a mess on lawns, sidewalks, and decks. Female ash trees produce numerous seeds, and the seedlings can become weedy; male seedless cultivars are available. Trees may bleed if pruned in spring.

Cultivars 'Bergeson' is a hardy cultivar with an upright, rounded crown. It originated in a nursery in Fertile, Minnesota. Zone 2. 'Marshall's Seedless' is supposedly seedless with good green foliage color. 'Prairie Dome' is seedless with a globe-shaped crown only growing to 40 feet. 'Prairie Spire' was selected for its upright pyramidal shape; it grows 55 feet tall. 'Summit' is a Minnesota introduction selected for its strong central leader. All other cultivars zone 3.

Other Species *F. americana* (white ash) is native to well-drained upland soils in the east-central and southeastern parts of the state. It grows up to 95 feet tall and fall color ranges from yellow to a distinctive maroon to deep purple. It is not as hardy (zone 4, protected spots in zone 3) or adaptable as green ash, but it is a handsome tree where its large size can be accommodated. It does best in rich, moist soils in full sun and it will suffer during dry periods. 'Autumn Blaze' is a Canadian introduction hardy into lower portions of zone 3 and only growing to 60 feet tall. 'Autumn Purple' is a male selection with deep purple fall color suitable for zone 4.

F. nigra (black ash) is native to moist deciduous and coniferous forests and wetlands in all but the southwestern part of Minnesota. It is slower growing and has a more irregular form, sometimes leaning or crooked. It is a hardy species, into zone 2, and it tolerates wet soils. It is not as susceptible to the insect and disease problems that plague green and white ash. 'Fallgold' is a seedless Minnesota introduction that turns yellow early in fall and is more adaptable to landscape situations, growing to only about 50 feet in height, but is only hardy into zone 3.

Fraxinus pennsylvanica

Gleditsia triacanthos
Honey locust
Zone 4

Native Habitat Moist or rich soils in the far southeastern corner of Minnesota.

Size 35 to 60 feet tall, 25 to 40 feet wide

Description Honey locust is a fine-textured, medium-sized, open-crowned tree with an airy look. The compound leaves are bright green, with autumn color an inconsistent yellow to yellow green. Sharp, shiny thorns up to 4 inches long appear on one-year-old wood and remain for many years. The fruit is a 10- to 18-inch, flat, dark brown pod that becomes black and twisted as it ripens.

Landscape Use This tree is best used in windbreaks and hedges in southern Minnesota. The sharp thorns and large seedpods restrict its use as a landscape plant, but there are several thornless selections that make good specimen and street trees. These selections are a good choice where you want light, dappled shade so you can grow other plants beneath the canopy. The small leaflets fall early in autumn and do not require raking. The flowers are a nectar source for bees, and the large seedpods provide food for wildlife.

Site Honey locust withstands a wide range of conditions, but it does best on rich, moist soils. It is tolerant of drought, high pH, road salt, and soil compaction. It does require full sun.

Culture Transplants easily and can be used in many urban situations. Honey locust is a thin-barked tree that must be protected while it is young. Surround trees with a circle of organic mulch to keep lawn mowers and weed whippers from damaging the trunk. The trunks of young trees should be wrapped in hardware cloth in winter to prevent damage from rodents. The thornless cultivars have been overplanted in recent years, and several insects and diseases have become serious problems, including a disabling trunk canker. The seedpods can be messy when they fall, and the main trunk may need some shaping when trees are young to ensure a good mature form.

Gleditsia triacanthos var. *inermis* 'Imperial'

Cultivars The species is rarely planted because of the thorns and heavy seed production. Many male seedless cultivars have been selected from the thornless variant, *G. triacanthos* var. *inermis*. 'Imperial' is more spreading than upright and often lacks a single central trunk. It grows to about 35 feet tall and produces few, if any, seeds. 'Shademaster' has a strong central trunk and ascending branches. 'Skyline' has a more pyramidal form and bright gold fall color. 'Sunburst' has bright yellow new growth that turns green in summer. It is only hardy in the southern part of Minnesota and often suffers dieback after severe winters. All cultivars hardy in zone 4.

Gymnocladus dioica
Kentucky coffee tree
Zone 4; trial in zone 3

Native Habitat Deep, rich soils in river valleys widely scattered in southern Minnesota.

Size 45 to 80 feet tall, 50 feet wide

Description Kentucky coffee tree has a picturesque, open-spreading crown. Its remarkably large, compound, bluish green leaves appear late in spring. The greenish white flowers that also appear in late spring are somewhat fragrant and attractive to pollinating insects. Fall color is inconsistent, sometimes a good yellow. The 6- to 10-inch-long, flat seedpods, deeply furrowed bark, and sparse branching add winter interest.

Landscape Use This is an excellent shade and street tree that deserves to be planted more. It is a good replacement for the disease-prone honey locust. The leaves cast very light shade (also known as filtered shade or dappled shade) that permits shade-tolerant turf grasses and partial-shade perennials to grow underneath. It is tolerant of city conditions and offers good winter interest. Leaves appear late in spring and drop soon after first fall frost, so it's a good choice where you want sunlight in spring and fall.

Site Grows best in evenly moist, rich soil, but is tolerant of alkaline soils and drought. Needs full sun.

Culture Kentucky coffee tree is slightly difficult to transplant because of the deep taproot. Plant smaller balled-and-burlapped or container-grown specimens in early spring. It is a moderate to slow grower. Encourage faster growth by fertilizing young trees in spring and providing supplemental water during dry periods. The large seedpods usually fall sporadically over a long period, but some people still consider them a litter problem. They are also toxic, so children should be discouraged from playing

Gymnocladus dioica

with them. Seed-free male selections are available. It is free of insect and disease problems.

Cultivars 'Stately Manor' is a University of Minnesota release with a more upright growth habit and a narrower crown. It's a male selection that does not produce seeds. 'Espresso' is a fruitless male selection with upward-arching branches in a vaselike form. Prairie Titan® is an upright-spreading male selection with good summer leaf color and interesting winter architecture. All cultivars reliably hardy in zone 4.

Juglans nigra
Black walnut
Zone 4; trial in zone 3

Native Habitat Rich bottomlands and moist, fertile hillsides in southern Minnesota.

Size 50 to 80 feet tall, 50 feet wide

Description Black walnut has a well-formed trunk devoid of branches halfway to two-thirds from the ground. The bark is thick and dark brown with deep fissures. The pinnately compound leaves have fourteen to twenty-two leaflets and reach up to 24 inches in length, usually turning yellow in autumn. The distinctive, large, round nuts are borne singly or in pairs enclosed in a solid green husk that does not open even after ripening. The nuts are edible, if you can get the husks open.

Landscape Use Black walnut is most attractive when it is grown as a specimen tree. It is best suited to large sites in southern Minnesota. Avoid planting it near sidewalks and streets where falling nuts can be a hazard and are messy. The nuts are a popular food source for squirrels and other wildlife, as well as humans.

Site Prefers full sun in deep, neutral to slightly alkaline soils but tolerates drier soil.

Culture The extensive taproot on older trees makes it difficult to transplant. Plant smaller balled-and-burlapped or container-grown specimens in early spring. Seeds can also be sown in place, provided some protection is given against hungry squirrels. There are no serious insect or disease problems, but some caterpillars use leaves as a food source. Black walnuts contain a chemical known as juglone, which is toxic to certain plants. The greatest quantities of juglone are found in the area immediately under the walnut tree, where roots are concentrated and decaying nut hulls and leaves accumulate. Symptoms on affected plants range from stunting, yellowing, and partial to total wilting to complete death. Tomatoes and potatoes are very sensitive to juglone; other plants known to be affected include rhododendrons, white pine, paper birch, eggplants, peppers, lilacs, and cotoneasters. If possible, locate gardens or landscape beds away from the root zone of trees or grow sensitive plants in containers under black walnuts to avoid root contact with the soil. Native plants usually not affected by juglone include wild bergamot, bloodroot, violets, cinnamon fern, Jacob's ladder, bellworts, trilliums, sensitive fern, and spiderworts.

Cultivars and Other Species

'Laciniata' is a rare form with leaflets that are fernlike and dissected with a fine texture. 'Thomas' and 'Weschcke' are two good nut-bearing cultivars for Minnesota. Zone 4.

J. cinerea (butternut) is native in rich woods and flood plains and drier rocky slopes in eastern Minnesota farther north than black walnut. Its susceptibility to butternut canker keeps it from being recommended as a landscape plant. However, native plants have been devastated by this disease and the species is on Minnesota's special concern list, so it is worth preserving if you have one on your property. It grows 40 to 65 feet tall with a more rounded crown than black walnut. The nuts are edible. It is hardier, well into zone 3.

Juglans nigra nut

Juglans nigra

Larix laricina
Tamarack
Zone 2

Native Habitat Wet soils and swamps and bogs in the coniferous forests of northern and east-central Minnesota, occasionally on drier sites.

Size 40 to 70 feet tall, 20 to 40 feet wide

Description Tamarack is the only deciduous conifer native to Minnesota. It has a narrow pyramidal form with short, horizontal branches. The bark is rough with thin, reddish brown scales. The leaves are needles that are flat, soft, slender, and about 1 inch long, borne in clusters on spurlike branches. They are bright green in spring and turn an attractive yellow-gold color in September to October just before falling.

Landscape Use Tamarack's landscape use is limited by its large size, lack of heat tolerance, and falling needles in winter, which cause some people to think the tree has died. It is a good choice for shelterbelts and naturalizing in wet areas in rural landscapes on a site where the beautiful fall color can be enjoyed.

Site Prefers full sun and moist soils, but it's surprisingly tolerant of drier, upland sites. It does not tolerate pollution, heat, or shade.

Culture Transplant when dormant in early spring into soil with sufficient moisture. Water young trees as needed to maintain soil moisture during the growing season and into early winter. An organic mulch of pine needles or shredded pine bark will help maintain soil moisture and soil acidity. Trees are subject to injury from larch case bearer and larch sawfly, but these are rarely problems in landscape situations.

Cultivars Several rare dwarf cultivars are available from specialty nurseries. They can be used as specimen plants or in the shrub border. 'Deborah Waxman' reaches only 4 feet in height. 'Lanark' grows low and wide. 'Newport Beauty' rarely exceeds 2 feet tall and wide. They are all hardy in zone 4, but can be tried in zone 3.

'Deborah Waxman'

Larix laricina

Ostrya virginiana
Ironwood
Zone 4; trial in zone 3

Native Habitat A variety of soils throughout Minnesota but scattered or absent along the western border.

Size 20 to 40 feet tall, 20 to 30 feet wide

Description Ironwood is a handsome, small- to medium-sized tree with many horizontal or drooping branches and a rounded outline. The dark green, birchlike leaves turn a mild yellow color in fall. The straight trunk and limbs are covered with interesting shaggy, gray bark. The seeds ripen in flattened, papery pods that are strung together like fish scales or hops and turn from light green to gray in late summer.

Landscape Use This tough tree is well suited to smaller city landscapes and tight spaces where most shade trees would grow too large. It can be grown as a clump tree. Its clean, disease-free foliage will provide good summer shade—dense if grown in sun and more open when grown in light shade. The wood is strong and resistant to ice and wind damage. The seedpods are eaten by birds or slowly disintegrate, so there is no litter problem. Its main drawback for use is its slow growth rate, but this can be somewhat overcome if it's given supplemental water and fertilizer when young. Use it as an understory tree in woodland gardens, where the slow growth is advantageous. Ironwood is not tolerant of salt or compacted soil, so it's not a good street tree.

Site Prefers a cool, moist, well-drained, slightly acidic soil, but adapts to a wide range of landscape situations provided the soil is acidic and not waterlogged. It tolerates full sun, partial shade, and even heavy shade.

Culture Transplant balled-and-burlapped or container-grown trees in early spring, using a northern seed source. Trees are somewhat slow to establish after transplanting. Regular watering and fertilizing when trees are

Ostrya virginiana

young will help plants come out of transplant shock quicker and become established. Apply a 2- to 4-inch layer of organic mulch such as pine needles, shredded oak leaves, or shredded pine bark to keep the soil evenly moist and maintain soil acidity. Prune in winter to remove dead or damaged branches and to shape trees, if needed. Trees may bleed sap in late winter. No serious insect or disease problems.

Prunus serotina
Black cherry
Zone 3

Native Habitat A wide variety of soils in the southern half of Minnesota but absent near the western border.

Size 50 to 75 feet tall, 40 to 50 feet wide

Description Black cherry has a uniformly thick trunk, often tilted or bent, and an open, rounded crown. The narrow, oval, shiny, green leaves develop a burgundy red to yellowish color in fall. The attractive, dark reddish brown bark has long, horizontal lenticels. On older trees, the bark becomes heavily textured, chunky, and beautiful. Attractive, ½-inch, white flowers appear in hanging clusters in May. The fruit is a green cherry, which turns red to dark blue or black and hangs in pendant clusters.

Landscape Use Although highly ornamental, black cherry is rarely planted in the landscape because of its suscepti-

Prunus serotina

bility to black knot. It is definitely worth preserving if you have one on your property, however, and it's a good choice for attracting wildlife and for naturalizing. When it survives in the landscape, it makes a handsome shade or specimen tree. The fruits are tart but edible and are a favorite of birds.

Site Will grow on almost any soil in full sun to part shade, but best growth is on deep, moist, slightly acidic, fertile soils in full sun.

Culture Moderately difficult to transplant because of the taproot. Move in early spring. Keep trees vigorous with ample water and soil fertility to help avoid pest problems. The occasional fruits that go uneaten by birds and drop to the ground may sprout new trees that can be weedy. Black knot is a serious fungal problem that can kill trees at a young age. It appears as a swollen black area on the branches. Cut out and destroy infected branches, going back about 4 inches below the knot.

Quercus alba
White oak
Zone 4; trial in zone 3

Native Habitat Abundant on heavy, well-drained, acidic soils in southeastern Minnesota, often forming woodlands almost to the exclusion of other trees.

Size 50 to 75 feet tall, 40 to 60 feet wide

Description White oak has a broad, rounded top with irregularly spreading limbs. The leaves, which don't fully unfold until mid-May, are deeply divided into five to nine fingerlike, rounded lobes. They are bright green in summer, turning deep red or violet-purple in fall.

Landscape Use A well-grown oak is an asset in any landscape, usually greatly increasing property value. Their stately presence is felt year round. The leaves offer summer shade, fall color, and often persist into winter for added interest. The tree's strong, bold silhouette also offers winter interest. The acorns are a valuable fall food source for many wildlife species. Do not be put off by oaks' reputation of slow growth or susceptibility to oak wilt, which is not usually a problem in landscape situations.

Site Oaks grow best in fertile, acidic, heavy soil in full sun. Most species will tolerate lighter, sandier soils and dry conditions, however.

Culture Oaks are somewhat difficult to transplant, making them hard to find in the nursery trade. Root pruning and container growing have made them more available and easier to transplant. Seedlings are fast growing, so don't hesitate to plant an acorn where you want a tree. Plant it in fall (with protection from digging squirrels) or store it in moist sand in a cool place over winter and plant in spring. Oak roots are sensitive to changes in soil level. Even a small change can kill roots, so avoid adding or removing soil around trees. Oak wilt, a fungal disease, can kill mature oaks; red oaks are more susceptible than white oaks. Oak wilt enters trees via insects through wounds caused by pruning or by root damage during construction. The fungus can spread to nearby trees by root grafts. Oak wilt is mainly a problem on native stands; usually landscape specimens are isolated enough not to be at risk. Do not prune while insects are most active, from April 1 to July 1.

Other Species *Q. bicolor* (swamp white oak) is found on wetter sites scattered in southeastern Minnesota. It grows 40 to 65 feet tall and has an open, rounded crown. It is easier to transplant than white oak and makes a handsome specimen or shade tree. It is also faster growing and more tolerant of tougher sites. Fall color is brown. Zone 4.

Q. ellipsoidalis (northern pin oak) is found on drier, sandier sites mainly in the eastern half of Minnesota, with some scattered populations in the west. It is smaller, growing to only 45 to 65 feet. Fall color is a good deep red to reddish brown. It will suffer from chlorosis on soils with too high of a pH, but this is easily corrected by adding iron to the soil and lowering the pH. It is susceptible to oak wilt. Zone 3.

Q. macrocarpa (bur oak) is native throughout Minnesota, extending into the western prairies. The leaves of this white-oak type are shaped like a bass fiddle. It grows 50 to 80 feet tall and makes an impressive specimen in a savanna setting. It is adaptable to various soils and more tolerant of city conditions than other oaks, but it can be difficult to transplant. It is too large for most city landscapes, but it is a good choice for rural and large suburban sites. Zone 3.

Q. rubra (red oak) is native to the eastern half of Minnesota, with some scattered populations in the west. It grows 50 to 80 feet tall and has pointed leaves. It is faster growing than other oaks and easier to transplant. The dark green leaves turn to shades of red in fall and persist into winter. It prefers sandy loam soils that are well drained and acidic; it will get iron chlorosis in high pH soils. It is tolerant of city conditions but is very susceptible to oak wilt. If you have a red oak on your property, it is definitely worth some effort to keep it healthy. Zone 4; trial in zone 3.

Quercus macrocarpa

Sorbus americana
American mountain ash
Zone 2

Native Habitat Cool, moist sites scattered in far northeastern Minnesota mainly along swamp edges.

Size 20 to 30 feet tall, 15 to 25 feet wide

Description American mountain ash is a small tree, often more shrublike with multiple trunks. The spreading, slender branches create a narrow, round-topped crown. Bark is light gray and smooth, with small platelike scales. Leaves are pinnately compound with thirteen to seventeen leaflets, bright green above, turning bright yellow in fall. Creamy white flowers appear in flat clusters in May and early June. The showy scarlet red berries are ornamental in September and October, especially in contrast to the fall leaf color.

Landscape Use This is a good single-trunked tree or a multistemmed, vase-shaped shrub. The fine-textured leaves cast light shade. It makes a nice specimen plant on a small site and can be used as a middle-layer shrub under the canopy of larger trees or massed for screening or a windbreak. Birds love the berries, which are usually stripped from the tree before winter.

Site Does best in acidic, moist locations in full sun, but tolerates drier areas and thinner soils if well mulched and watered. Plan this understory tree where it will be sheltered from afternoon sun in winter.

Culture American mountain ash is hardier and not as susceptible to fire blight as the nonnative European mountain ash (*S. aucuparia*). To reduce chances of infection, avoid wounding the trunk and prune only while dormant. Young trees should be protected against sunscald and rodent feeding in winter. Yellow-bellied sapsuckers often drill rows of horizontal holes to suck the sap, and this can do enough damage to kill trees.

Other Species *S. decora* (showy mountain ash) grows native over a wider

Sorbus decora

range in northeastern Minnesota. The leaflets are broader and darker green than *S. americana*. It blooms and fruits a week or two later, and the fruits are slightly larger and deeper red. It is moderately shade tolerant and slow growing. Zone 2.

Tilia americana
Basswood
Zone 3; protected spots in zone 2

Tilia americana 'Fastigiata'

Native Habitat Throughout Minnesota, usually on rich soils.

Size 50 to 90 feet tall, 40 to 50 feet wide

Description This large tree has a dense, rounded crown and large, heart-shaped, dark green leaves. Autumn color is a dull green to slightly yellow green. The light gray, smooth bark becomes dark gray with age. The cream-colored flowers appear in June and July; they are fragrant and attractive to bees. The fruit is a rounded nutlike drupe that hangs onto the tree long into winter.

Landscape Use Basswood is a trouble-free, hardy tree that makes a good specimen or shade tree if you have room for its large size. It casts deep shade that is difficult to grow other plants under. It's not tolerant of soil compaction or air pollution, so it's not well suited to street-tree use.

Site Prefers moist, fertile, well-drained soils but will grow on drier, heavier soils in full sun or part shade.

Culture Basswood is easily transplanted in early spring. It is moderately fast growing and long-lived. Wrap young trees in winter to prevent sunscald. Prune out sprouts that develop at the base of the tree. Few insects or diseases bother it.

Cultivars Several cultivars have been selected for their smaller size, but most are not quite as hardy as the species. 'Bailyard' (Frontyard®) has a broadly pyramidal shape and good yellow fall color; it grows to 70 feet in height. 'Boulevard' grows 60 feet tall and has a more columnar shape. 'Fastigiata' is more pyramidal and grows to 50 feet tall and about 35 feet wide. 'Sentry' has an upright-branching habit. The cultivars are hardy in zone 4 and can be tried in zone 3.

Additional Native Deciduous Trees

***Crataegus mollis* (downy hawthorn)** is native to dry soils throughout Minnesota. It is a rounded tree growing 20 to 30 feet tall with several ornamental qualities. The leaves are thick and shiny dark green, and the fragrant flowers are white, sometimes pink, in clusters. The apple-like red fruits are showy and eaten by birds and animals. Its use is limited by its susceptibility to hawthorn rust and the 2-inch thorns that cover the branches. It can be used in naturalized plantings and shelterbelts. Birds nest in the branches for protection. Zone 3.

***Morus rubra* (red mulberry)** is native to moist soils and floodplains in a few spots in the southeastern corner of Minnesota. It grows to 50 feet tall with a single trunk and spreading branches. The shiny green leaves turn yellow in fall. The prolific fruits are green berries turning red to black. The fruits are a food source for birds and other wildlife, and they can be used in jams, jellies, and pies. Trees often become weedy in the landscape after birds deposit the seeds, and the fruits are messy when they fall. Protected sites in zone 4.

***Populus tremuloides* (quaking aspen)** inhabits wet or dry soils throughout Minnesota. It is fast growing to about 65 feet tall. It has good golden yellow fall color and the leaves make a wonderful sound when they flutter in the slightest breeze.

However, it is short-lived, weedy, and suckers, limiting its landscape use to naturalistic plantings and shelterbelts. Zone 3.

P. grandidentata (big-toothed aspen) is native on moist soils in the eastern two-thirds of Minnesota and suffers from the same problems. Zone 3.

P. deltoides var. *occidentalis* (eastern cottonwood) is native to wet soils along streams, rivers, and lakes throughout Minnesota, except for the northeastern quarter. It is large, growing 70 to 100 feet, and the seeds make a cottony mess in spring. Zone 3.

Populus tremuloides

***Salix nigra* (black willow)** is native to wet soils, stream banks, and wetlands in eastern Minnesota. It grows 40 to 60 feet tall or more with a single crooked trunk, often forked, and a narrow, irregular crown. Leaves are shiny green, turning light yellow in fall. It is fast growing and short-lived, and the brittle wood can become messy. Consider it for use as a lawn tree or for screening in an area with moist soil. Zone 3.

S. amygdaloides (peach-leaved willow), native throughout Minnesota, is similar to *S. nigra*. Zone 3.

***Ulmus americana* (American elm)** is native to moist soils throughout Minnesota, most abundantly on rich bottomlands in the southern half of the state. It grows 70 to 100 feet tall and 50 to 70 feet wide. The vase-shaped form and arching branches of American elm are quite distinctive, and the dark green leaves turn yellow in fall. Once a favorite because of their urban tolerance, fast growth, and unique vase-shaped silhouette, landscape planting of American elms is not recommended at this time due to the widespread Dutch elm disease.

U. rubra (red elm) is native to river bottoms and lowlands mainly in southern Minnesota. It grows 50 to 70 feet tall. It is less susceptible to Dutch elm disease, but is not nearly as ornamental and becomes weedy in the landscape. It can be used for shelterbelts. Zone 3.

Deciduous Shrubs and Small Trees

Shrubs and small trees fulfill many important roles in the landscape. They are the basis for most foundation plantings and hedges. They fill out mixed borders, providing interest all year long and especially in winter when flowers and groundcovers have little to offer. Many shrubs and small trees have fruits that provide winter interest as well as attract birds and other wildlife. In shade gardens, they provide that middle tier that is often forgotten in the home landscape. Shrubs are valuable for hedges and for massing. Small trees are important for adding height to entry and patio gardens, and many make nice lawn trees in small yards. Most large shrubs can be pruned into small trees, adding another dimension to their landscape use.

To use shrubs and small trees intelligently you must know their mature size and under what conditions they grow best. Consider hardiness, light requirements, growth rate, and seasonal aspects like flowers, fruits, fall color, and winter appearance. It is best to select a plant based on your existing soil, but it is fairly easy to change soil conditions if necessary.

Planting

Deciduous plants are best planted in spring. If you have the opportunity to move a native tree or shrub, move it in early spring. Bare-root plants must be planted in early spring as soon as the soil can be worked. Container plants can be planted anytime but the hottest days of summer, during July and August. Spring planting is still the best, however.

Before planting, amend the soil with a good amount of organic matter such as compost, peat moss, or well-

Cornus alternifolia is readily adaptable to landscape use, where it can be used as a large shrub or pruned as an interesting small tree.

rotted manure. Mix this thoroughly with the planting-hole soil. Place the shrub or small tree at the same depth it was growing in the container. Bare-root plants should be planted so that the crown is level with the ground level. Newly planted shrubs and small trees should not need additional fertilizer. Do not stake trees unless they are on an extremely windy, open site. It is a good idea to surround all newly planted woody plants with a ring of organic mulch 2 to 4 inches thick. Good mulches are wood chips, shredded bark, and pine needles. Replenish the mulch as needed throughout the growing season.

Care

The first two or three years after planting, make sure the soil is evenly moist from spring until the ground freezes in fall. Once established, many shrubs can tolerate some dry periods, but don't hesitate to water as needed, especially in sandy soils. Always saturate the soil thoroughly with each watering. Most woody plants will benefit from a spring application of fertilizer. Spread a layer of rotted manure or compost around each plant or use Milorganite or fish emulsion. If possible, allow leaves to fall and decay under shrubs to return nutrients to the soil. Keep weeds pulled or smother them with organic mulch. Do not use rock or black plastic as mulch.

Maintenance needs differ with each species. Most small trees will benefit from some selective pruning while they are young to develop a good shape. The best time to do this is usually while they are dormant: you can see their silhouette better, and it reduces the chances of disease and insect problems. Remove branches that have narrow-angled crotches as these branches are weaker than wide branches and are likely to split as the tree gets older. Shrubs are usually pruned according to their flowering time. In general, spring-flowering shrubs, which set their flower buds in summer, should be pruned right after they have finished flowering. Summer-flowering shrubs, which develop flower buds in spring, are usually best pruned in early spring. Remove any dead, damaged, or diseased parts at any time of year, cutting back to just above a healthy, outfacing bud. Shrubs that have become overgrown will benefit from renewal pruning. To do this, cut the plants back to 4 to 6 inches from the ground in early spring. Surround them with a layer of compost or rotted manure and water them well to help them recover.

Possible Problems

Most insect and disease problems are purely cosmetic and don't really threaten the life of shrubs or small trees. The best defense against insect and disease problems is a vigorously growing plant. By choosing the appropriate plant for your site conditions and providing it with the necessary nutrients and ample water, most pest problems will not become serious. Powdery mildew can affect some shrubs, especially when they are growing close together. Snip out the infected areas and dispose of the foliage. Thin out plants to allow for better air circulation and drying of foliage. Leaf spots may leave plants somewhat defoliated by late summer, but this is rarely a serious problem. Rake up leaves to reduce the source of inoculum for next year. Aphids often feed on young leaves in hot, dry weather. Dislodge them by spraying the foliage daily with a hard spray from the garden hose. Young trees should be protected from winter rodent feeding, which can kill a tree. The best way to do this is to surround the trunk below the first branch with hardware cloth.

Best Native Deciduous Shrubs and Small Trees for Landscape Use

Amelanchier species (serviceberries)
Cornus species (dogwoods)
Dirca palustris (leatherwood)
Ilex verticillata (winterberry)
Physocarpus opulifolius cultivars (ninebark)
Potentilla fruticosa cultivars (shrubby cinquefoil)
Prunus nigra 'Princess Kay' (Canada plum)
Viburnum species (viburnums)

Amelanchier laevis
Smooth juneberry, Alleghany serviceberry
Zone 3

Native Habitat Wood edges, moist hillsides, and ravines mainly in the eastern half of Minnesota.

Size 25 to 30 feet tall, 15 to 25 feet wide

Description Juneberry is a single or multi-stemmed small tree or large shrub. Leaves are purple in early spring, turning green in summer and red orange in fall. The delicate white flowers start blooming in late April with the just-emerging reddish spring foliage. Deep purple-red fruits appear in summer. The bark is silvery gray.

Landscape Use Use this shade-tolerant shrub as an understory plant in deciduous woodland plantings, shade gardens, and shrub borders. The smooth bark is attractive in early spring in wildflower gardens. It can be pruned as a specimen tree near patios and in entry gardens. All *Amelanchier* species are excellent for naturalized settings, where their fruits attract wildlife. The fruits can also be used for jams, jellies, pies, and wine—if you can keep them from the birds.

Site Prefers moist, well-drained, slightly acidic soil high in organic matter in part shade to sun.

Culture All *Amelanchier* species are easily transplanted and adaptable to culture. Most will tolerate dry, poor conditions once established. Remove some of the older stems each year in late winter to keep plants vigorous and producing more fruits. Renew overgrown shrubs with hard pruning.

Prune out suckers if desired; they can be dug up and transplanted in early spring.

Other Species *A. alnifolia* (Saskatoon juneberry) is native to stream banks, thickets, and slopes in all but southern Minnesota. It grows 3 to 12 feet tall and 6 to 9 feet wide. It has abundant, sweet, purplish-black fruits, making it a good choice for edible landscaping. Shrubs are attractive in bloom and in fruit, but they do sucker profusely. 'Regent' is a popular cultivar growing about 5 feet tall with good flowering, foliage, and fruiting. Zone 2.

A. arborea (downy serviceberry) is native to dry soils, hillsides, and forest edges along the eastern edge of Minnesota. It grows 20 to 25 feet tall and 15 to 20 feet wide. It has multiple narrow trunks and a rounded crown. Zone 3.

A. × grandiflora (apple serviceberry) is a naturally occurring hybrid of *A. arborea* and *A. laevis*. It is an excellent small clump or single-trunked tree growing to 25 feet. Several cultivars have been selected for their good fall color and interesting growth habit. Good choices for Minnesota include 'Autumn Brilliance', 'Robin Hill', 'Princess Diana', and 'Strata'. Zone 4.

A. stolonifera (running serviceberry) is native to acidic, sandy, rocky outcroppings in far northeastern Minnesota. It grows 1 to 6 feet tall and 3 to 10 feet wide. Its spreading growth habit makes it suitable for consideration as drought-tolerant groundcover. Zone 3.

Amelanchier × grandiflora 'Autumn Brilliance'

Amelanchier alnifolia 'Regent'

Amorpha canescens
Leadplant
Zone 3

Native Habitat Loamy or sandy, moist to dry open woods, savannas, and prairies in all but northeastern Minnesota.

Size 1½ to 4 feet tall, 2 to 3 feet wide

Description Leadplant is a semi-woody, loose shrub with numerous spiked clusters of tightly packed, small, purple flowers June into August. The pinnately compound leaves and stems are covered in dense, woolly, gray hairs, giving plants a grayish color. It is one of the few woody species common on native prairies. It is especially sensitive to disturbance; hence, its presence indicates that a prairie has not been cultivated.

Landscape Use Use leadplant in prairie gardens, mixed borders, and shrub plantings, where the foliage is attractive all season. It is also good for xeriscaping.

Site Prefers dry, average, well-drained soil in full sun, light shade at most.

Culture Leadplant requires a nitrogen-fixing soil bacterium. If it is not present, use a commercial soil inoculant when planting. Mulch plants the first winter to prevent frost heaving. Its deep-branching taproot makes leadplant difficult to transplant but helps it survive fire, which improves growth and bloom of prairie plants. It is a favorite food of deer. Plants can be rejuvenated in spring by cutting them to the ground or pruning out old wood. It is often treated more like an herbaceous perennial, especially when grown in a mixed border.

Other Species *A. fruticosa* (false indigo) is native to moist woods and stream banks in southern and western Minnesota. It is a thicket shrub growing about 12 feet tall with spikes of dark purple to pale blue flowers in May and June and green leaves. It requires moist soil and is a good choice for naturalistic plantings near water. Zone 3.

A. nana (dwarf wild indigo, fragrant false indigo) is native to dry soils in western and southern Minnesota. It only grows 1 to 2 feet tall and is drought tolerant. Zone 3.

Amorpha canescens

Amorpha fruticosa

Arctostaphylos uva-ursi
Bearberry
Zone 3

Native Habitat Sandy, shallow, acidic soils on rock outcrops in northern and eastern Minnesota.

Size Stems up to 8 feet long; branches 6 to 12 inches tall

Description Bearberry is a prostrate evergreen shrub with trailing stems. The leaves are dark green and leathery, often turning purplish in fall. Small clusters of tiny white or pink, urn-shaped flowers appear in late April and May. The showy clusters of bright red berries replace the flowers in fall and persist through winter, even into early spring.

Landscape Use This is an excellent groundcover for dry sites in sun to part shade. It will trail over rocks and down banks. The ornamental fruits are showy in fall and winter and are eventually eaten by songbirds and game birds. It is well suited to coarse-textured soils low in nutrients and can be grown on gentle to steep slopes for erosion control.

Site Prefers dry, poor, acidic soil in full sun to light shade; grows best in full sun. Does poorly on neutral or slightly alkaline soils.

Culture Acidify and loosen soil before planting if needed. Bearberry is somewhat difficult to establish, but once it takes hold it's a tough plant. Transplant container-grown plants from a northern seed source in early spring. Pay careful attention to water the first two years or so, until the roots firmly take hold. A 1- to 2-inch layer of compost mulch will help conserve moisture and maintain soil acidity.

Arctostaphylos uva-ursi 'Massachusetts'

Aronia melanocarpa
Black chokeberry
Zone 3; trial in zone 2

Native Habitat Swamps, low areas, or open coniferous woods in acidic soils in east-central and northeastern Minnesota.

Size 4 to 6 feet tall, 3 to 6 feet wide

Description This upright, mounded shrub has dark green, glossy foliage in summer that turns a beautiful red in fall. White flowers produced in early May last for two weeks or more, followed by small black fruits that resemble tiny apples.

Landscape Use Black chokeberry is a tough, reliable shrub that offers year-round interest, but is especially showy in fall. Use it in shrub borders and naturalistic plantings. Lower branches often become sparse; hide this by using low-growing plants in front of black chokeberry. The fruits are low on the list for birds. They are ornamental enough for use in wreaths and floral arrangements. Although technically edible, the fruits are extremely tart and bitter and are not recommended for eating off the bush.

Site Grows in slightly acidic soil in both wet and dry conditions in full sun to light shade.

Culture Easy to transplant. Plants do best if left untrimmed. This shrub can be somewhat invasive when the suckers are allowed to remain. Suckers are easily controlled if desired.

Cultivar 'Autumn Magic' is an upright grower to 5 feet with good flower and fruit production and outstanding fall color. Zone 3.

Aronia melanocarpa 'Autumn Magic'

Ceanothus americanus var. pitcheri
New Jersey tea
Zone 4; trial in zone 3

Native Habitat Well-drained prairies, oak savannas, and woodland edges in southern and eastern Minnesota.

Size 2 to 3 feet tall, 2 to 3 feet wide

Description New Jersey tea is a low-growing, spreading, many-branched shrub with dark green leaves with prominent veins. The showy, creamy white flowers appear in upright, umbel-like clusters in midsummer when not many plants are in bloom. The dried seed capsules add interest in late summer. Fall color is a nonspectacular yellow green.

Landscape Use A durable small shrub once established, New Jersey tea can be used in mixed or shrub borders and foundation plantings. It makes a beautiful small hedge when planted 1 to 2 feet apart. The flowers attract bees, butterflies, and hummingbirds, and birds eat the seeds. It is a good cut or dried flower.

Site A well-drained sandy soil in sun or part shade. Will not tolerate wet soils.

Culture Young plants are easy to move, but older ones are hard to transplant because of the extensive taproot. An occasional hard pruning in late winter will help keep plants looking neat. New Jersey tea is often treated more like an herbaceous perennial, especially when grown in mixed borders. High humidity and heavy summer rainfall can be harmful to plants, especially if they are in heavy soils in partial shade.

Ceanothus americanus

Cephalanthus occidentalis
Buttonbush
Zone 4

Cephalanthus occidentalis

Native Habitat Swamps and along streams in southeastern Minnesota.

Size 3 to 8 feet or more, 3 to 6 feet wide

Description Buttonbush is a low-branched shrub with smooth, gray-green bark. It has attractive, medium green leaves that nicely set off the white flowers borne in a 1-inch globe with protruding pistils. It blooms when few other plants are in flower late July into August. Bright red fruits form in autumn.

Landscape Use Plants have an almost tropical look when the interesting flowers appear against the glossy green leaves. Flowers have a musky, sweet scent. Leaves cluster toward the outside of the canopy giving an open area underneath for planting. Buttonbush can be grown in standing shallow water and is good for naturalizing alongside ponds. The flowers are a nectar source for butterflies.

Site Moist or wet, fertile soils; intolerant of drought.

Culture Plants can be cut back to 6 inches each winter; they'll grow to 3 feet tall by midsummer and still flower on the new growth. Buttonbush is often treated more like an herbaceous perennial, especially when grown in mixed borders. It often dies back to the ground after a severe winter. Landscape plants should be well mulched to retain the necessary soil moisture.

Comptonia peregrina
Sweet fern
Zone 3; trial in zone 2

Comptonia peregrina

Native Habitat Dry, sandy, acidic soils in northeastern Minnesota.

Size 1 to 4 feet tall, 2 to 3 feet wide

Description Sweet fern is not a true fern, but rather a low-growing, rhizomatous shrub with dark green, 2- to 6-inch-long, fernlike leaves that have a sweet fragrance.

Landscape Use Sweet fern spreads mainly by rhizomes, forming thickets, and it is effective in controlling erosion on slopes and difficult sites. The fine-textured leaves are aromatic, and the curled, dried foliage remains all winter, adding interest. A carpet of sweet fern will give a landscape a definite woodsy feel. In the wild, it often grows beneath *Pinus banksiana* (Jack pine). Flickers eat the small brown fruits, and prairie chickens and sharp-tailed grouse use plants for nesting cover.

Site Prefers dry, acidic, sandy soils, but will survive on most garden soils as long as they are acidic.

Culture Acidify soil before planting, if necessary. Established plants are difficult to move, so start with young container-grown specimens for best success. Sweet fern fixes nitrogen in the soil, which helps it survive tough conditions and eliminates its need for additional fertilizer. It maintains its mature 3- to 4-foot height for a long time without pruning. Remove some of the older canes in spring to keep plants vigorous. Pest free.

Cornus alternifolia
Pagoda dogwood
Zone 3

Native Habitat Well-drained, moist, open woodlands in the eastern two-thirds of Minnesota.

Size 20 to 25 feet tall, 20 to 25 feet wide

Description This small understory tree or large shrub grows in horizontal tiers of branches giving it a layered appearance. The deep green leaves are heavily veined and turn reddish in fall. Small, creamy white, musky-scented flowers appear in 3- to 5-inch clusters late May to early June. The fruit is green and berrylike, turning white to blue to nearly black on red stalks.

Landscape Use Pagoda dogwood is an excellent landscape plant. It can be grown as a large multistemmed tree or easily pruned to an attractive specimen tree. It has a graceful, tiered branching pattern. Use it as an understory shrub in woodland gardens, shrub borders, or as an entry or patio tree. Birds love the fruits.

Site Pagoda dogwood does best in cool, moist, slightly acidic soils in partial shade. It thrives in mulched landscape beds, but will not do well as a lawn specimen.

Culture Available balled-and-burlapped and container-grown for spring planting. Acidify soil before planting, if necessary. A little selective pruning in winter to encourage the horizontal branching habit will pay off in a lovely landscape specimen. Plants may self-seed in surrounding garden beds, but they are easily pulled out or transplanted. Plants grown in full sun should be well mulched and watered. Dogwoods are susceptible to powdery mildew and leaf spot in wet years. These diseases result in unattractive foliage in late summer and early fall, but are not serious enough to kill plants. Reduce chances of infection by keeping plants well watered and pruned to increase air circulation. Cankers causing stem dieback are usually caused by lack of water. Prune out infected stems several inches below the canker or at ground level.

Cultivars and Other Species Golden Shadows™ ('W. Stackman') is a new var-

Cornus alternifolia

Cornus racemosa

Cornus stolonifera

iegated selection with leaves displaying a broad lime-green to chartreuse central zone surrounded by a well-defined, iridescent golden-yellow margin. Zone 3.

C. racemosa (gray dogwood) is native in open woods throughout most of Minnesota. It grows 8 to 12 feet tall and up to 10 feet wide. The attractive gray stems support creamy white flowers in May, followed by showy white fruits borne on red pedicels in late summer. Birds eat the fruits, but the showy red pedicels persist to contrast nicely with snow. It grows best in moist, cool soils in full sun, but is tolerant of dry conditions and partial shade. It spreads slowly by underground stems and is excellent for naturalizing at the edges of woods or for hedging, but it is also successfully grown as a specimen tree with persistent pruning. Stick with northern-grown strains to ensure hardiness. Zone 4; trial in zone 3.

C. rugosa (round-leaved dogwood) is native to woodlands in all but southwestern and western Minnesota. It grows 10 feet tall in dense shade and is useful as an understory shrub in woodland gardens. Plants may be difficult to locate. Zone 2.

C. stolonifera (red-osier dogwood) is native to swamps, low meadows, and forest openings and margins throughout Minnesota. It is an attractive 10- to 12-foot landscape shrub, with deep red stems and twigs that are showy in winter, creamy white flowers in spring followed by attractive white fruits, and maroon-colored fall leaves. Plant it in shrub borders, foundation plantings, or outside a winter window with evergreens as a backdrop to set off the color. Once established, it is drought tolerant and low on the list of deer favorites. It spreads by layering when the lower stems touch or lie along the ground. It can form dense thickets in the right conditions, and can be used as a hedge. Younger stems have the brightest color, so prune out oldest stems each spring to encourage new growth. Overgrown plants can be cut back to about 6 inches in spring. It is the same plant as *C. sericea*. 'Cardinal' is a University of Minnesota introduction with bright cherry-red stems. 'Isanti' is another Minnesota introduction with a compact growth habit to about 6 feet tall and good stem color. Zone 3.

Corylus americana
American hazelnut
Zone 3

Native Habitat Both dry and moist soils in thickets in all but southwestern Minnesota.

Size 10 feet tall, 6 feet wide

Description This multistemmed, rounded shrub has dark green leaves that are slightly hairy above and softly hairy beneath. Fall color is sometimes yellowish. The small but interesting dangling catkins appear in early to mid-April and are followed by edible nuts maturing in fall.

Landscape Use American hazelnut has attractive summer foliage and interesting fall fruits, but its large size and coarse growth habit limit its landscape use to screening, naturalistic plantings, or large shrub borders. The nuts are favorites of squirrels.

Site Well-drained, loamy soil in full sun or light shade.

Culture Easy to transplant and grow. Suckers, which form freely from the base, can be pruned out at any time. Regular pruning helps keep plants neater. No serious insect or disease problems.

Other Species *C. cornuta* ssp. *cornuta* (beaked hazelnut) is native to the same areas of Minnesota. It is similar except it has a longer beak on its fruits and is slightly hardier, into Zone 2.

Corylus americana fruits

Corylus americana

Diervilla lonicera
Bush honeysuckle
Zone 3; trial in zone 2

Native Habitat Exposed, rocky sites and dry to mesic, well-drained soils mainly in the eastern half of Minnesota.

Size 2 to 3 feet tall, 2 to 3 feet wide

Description This low, suckering shrub has attractive glossy, green leaves tinged with red, and good orange to red fall color. Yellow, trumpet-shaped flowers appear in late June into July and are pretty but not exceptionally showy. The fruit is a dried capsule that is somewhat showy. Plants send up stems from underground rhizomes.

Landscape Use Bush honeysuckle is a rugged, pest-free plant with a long bloom time and attractive foliage. Use it as groundcover or bank cover. It is a good low-maintenance choice for massing and naturalizing on tough sites, and is one of the few shrubs able to tolerate the dry shade under large trees.

Site Adaptable, but prefers organic, well-drained soils in sun or part shade; tolerates high pH, compacted soils, and windy conditions.

Culture Easy to transplant. Give plants room to expand to their full size. Keep plants in bound by pulling suckers as needed. Plants can be cut back to the ground in spring to improve appearance. Plants rarely require fertilizing.

Diervilla lonicera

Dirca palustris

Dirca palustris
Leatherwood
Zone 2

Native Habitat Rich woodlands and swamp margins in the eastern half of Minnesota.

Size 3 to 6 feet tall, 4 to 6 feet wide

Description This deciduous shrub has attractive, smooth, grayish brown bark and flexible twigs that can become knobby, creating an interesting winter silhouette. Small yellow flowers appear in mid to late April just as the leaves are beginning to open. The gray-green leaves turn a soft yellow color in fall.

Landscape Use Leatherwood offers a subtle beauty to the landscape all year round. The early spring blooms and bright green foliage are an especially welcome sight, and the clean summer foliage, fall color, and winter shape add to its list of assets. It can be grown as a specimen, in shrub borders, or naturalized. Its shade tolerance makes it useful in woodland gardens.

Site Prefers moist, slightly acidic soil in part or even heavy shade. Plants in full sun may have bleached-out foliage.

Culture Acidify soil before planting, if necessary. Leatherwood stays dense and symmetrical in full sun, requiring little pruning or shaping. In shade, it is more irregular in shape and may benefit from light pruning after flowering. Plants grown in sun should be well watered and mulched. No pest problems.

Elaeagnus commutata

Elaeagnus commutata
Silverberry
Zone 2

Native Habitat Dry, limestone soils mainly in northwestern Minnesota.

Size 6 to 10 feet tall, 6 to 8 feet wide

Description This upright, stoloniferous shrub suckers freely to form large thickets. The beautiful pewter-silver foliage has an attractive sheen. The spicy-scented, small, yellow flowers appear in June. They are followed by green berries that turn orange and red and have a metallic sheen.

Landscape Use Though ornamental, silverberry's suckering habit reduces its landscape use. Use it for naturalizing or massing on hot, dry, windswept sites such as along parking areas or on south-facing slopes. The foliage is prized for dried flower arrangements. Birds eat the fruits.

Site Prefers a well-drained, low-fertility soil in full sun.

Culture Easily transplanted. Drought resistant once established. The suckering habit can be a problem. No pruning needed. No pests.

Euonymus atropurpurea
Wahoo
Zone 2

Native Habitat Rich, moist soils usually along streams, rivers, and floodplains in the southeastern quarter of Minnesota.

Size 20 to 25 feet tall, 10 to 12 feet wide

Description This large shrub or small tree has an interesting flat-topped, irregular crown. It is showiest in fall, when the dark green leaves turn a good reddish purple color and showy pink fruits open to reveal scarlet-orange seed coverings that ripen over a long period.

Landscape Use Wahoo can be pruned to a single- or multiple-trunked tree with an irregular crown and corky stems. With a minimal amount of grooming, it can be used as an understory shrub in woodland gardens, shrub borders, or as a specimen tree. Left alone, it is good for naturalizing and attracting birds. It should be used more in landscape situations.

Site Prefers a rich, moist, slightly acidic soil high in organic matter but will tolerate most landscape situations in full sun or part shade. Fall color is better in full sun, but it will grow in heavy shade.

Culture Wahoo requires consistent soil moisture. Mulch plants to keep soil moist and acidic. Prune in early spring to shape plants; tolerates heavy pruning. No serious insect or disease problems, but rabbits and deer may feed on the bark and leaves. It may be difficult to find in the nursery trade.

Euonymus atropurpurea

Hamamelis virginiana
Witch hazel
Zone 4; trial in zone 3

Native Habitat Sheltered ravines in hardwood forests in a few spots in the southeastern corner of Minnesota.

Size 10 to 12 feet tall, 10 to 12 feet wide

Description This deciduous shrub is the last shrub to bloom in autumn, producing golden-yellow flowers that are slightly fragrant. The flowers take a back seat to the wonderful golden-yellow fall leaf color. This open, multistemmed shrub has a tighter growth habit when grown in more light. Branches have a zigzag pattern.

Landscape Use Witch hazel's coarse texture limits its formal landscape use, but it's a good choice for the middle layer in woodland gardens. It is also suitable for shrub borders, where other plants can add interest until fall, when it shines. It can be pruned into a small, wide-spreading tree.

Site Prefers a uniformly moist soil in sun or part shade but is tolerant of dry, shady conditions.

Culture Fast-growing when young. Plants grown in full sun may show leaf scorch in summer. Rejuvenate overgrown shrubs with heavy pruning in early spring. Witch hazel is on Minnesota's special concern list, so do not dig plants from the wild and only buy nursery-propagated plants.

Hamamelis virginiana

Ilex verticillata var. *verticillata*
Winterberry
Zone 3; trial in zone 2

Native Habitat Acidic swamps and wet woods in eastern Minnesota.

Size 6 to 10 feet tall, 6 to 10 feet wide

Description Winterberry is a deciduous holly, one of the few hollies to lose its leaves in winter. Leaves vary from flat to shiny on the upper surface. Autumn color is not especially showy, but the bright red berries produced in tight clusters along stems are showy after the leaves drop and up until winter birds devour them.

Landscape Use From late fall through winter, winterberry steps into the spotlight by producing an outstanding display of bright red berries that persist on the branches even after the leaves have fallen. Plant it in groupings or mix with other plants that lack winter interest. It can be used for hedging. The red berries are extremely effective when contrasted against background snow or when reflected in nearby bodies of water.

Site Best performance is in full sun in acidic, organically enriched, moist to wet soils, but it is somewhat adaptable to soils that are occasionally dry. Chlorosis and stunting will occur in alkaline soils.

Culture Berries are only produced on female plants. Plant one male plant in close proximity to three to five female plants to ensure good pollination and subsequent fruit set. Plants in relatively dry soils will have better berry size (and subsequent ornamental appeal) with irrigation during dry periods in July and August. Young plants will grow faster with annual applications of an acidic fertilizer. Little pruning is needed.

Cultivars Several cultivars have been selected. Most are females that require a male pollinator to set fruit. 'Afterglow' features smaller, glossy green leaves and large orange-red berries maturing to orange. 'Cacapon' grows to 6 feet with

Ilex verticillata 'Compacta'

bright red fruits. 'Red Sprite' (also known as 'Compacta') is a popular dwarf maturing at only 3 to 4 feet tall. 'Shaver' has large clusters of red-orange berries. Use 'Jim Dandy' to pollinate all three. 'Winter Red' has a good growth habit and profuse bright red fruits that consistently persist into winter; 'Southern Gentleman' is a good pollinator.

Physocarpus opulifolius
Ninebark
Zone 2

Native Habitat Rocky soils along streams and lakes from the north shore of Lake Superior south along the eastern border into southeastern Minnesota.

Size 6 to 9 feet tall, 6 to 9 feet wide

Description This tough and hardy multi-stemmed shrub produces fast-growing shoots that arch out and away from the center. The five-petaled, white to pinkish flowers are grouped together in 2-inch, flat-topped clusters in June and July and are followed by reddish brown fruits that are somewhat showy. Older stems are covered with attractive, shaggy bark that sloughs off in long fibrous strips, but the foliage usually covers it. Foliage stays clean and attractive all season and can be shades of yellow or purple. Dried fruits can vary from brown to bright red, depending on soil and weather.

Landscape Use Use ninebark and its cultivars in hedges, foundation

plantings, and shrub borders. The foliage adds interest without becoming sickly looking like some other yellow-leaved plants. The species is good for screening and providing wildlife shelter. The flowers and foliage can be used in cut arrangements.

Site Full sun. Tolerates a wide range of soil types as long as they are well drained.

Culture Space hedge plants 2 to 3 feet apart. Ninebark, especially the cultivars, will benefit from a spring application of an organic fertilizer. Prune plants as needed to shape right after flowering. Overgrown plants can be cut back in late winter, but this shouldn't be done every year. Hedges should receive light pruning annually; they will have fewer blooms than unpruned plants. Hedge plants may suffer from powdery mildew in wet years, but otherwise these plants are trouble free.

Cultivars Several cultivars have been

Physocarpus opulifolius 'Nugget'

developed that are better suited to more-formal landscape situations. 'Dart's Gold' grows to a compact 4 to 6 feet tall and wide with good foliage color. Diabolo® has distinctive purple foliage that ages to a bronze shade. It can be pruned harshly each spring to promote vigorous shoots with large, highly colored leaves. 'Nugget' grows to only 6 feet tall and wide with greenish yellow, textured leaves. 'Snowfall' flowers more freely and has medium green leaves. Zone 2.

Potentilla fruticosa
Shrubby cinquefoil
Zone 2

Native Habitat Poor soils in sunny bogs and rocky outcroppings in north-western, northeastern, and east-central Minnesota.

Size 1 to 4 feet tall, 1 to 4 feet wide

Description Shrubby cinquefoil has yellow flowers that bloom from June to frost. Flowers of the many cultivars come in shades of white, yellow, and gold, which hold up well in summer heat, and orange, pink, and red, which require cooler summer temperatures to hold their color. The grayish green compound leaves are fine textured and attractive. Persistent seed heads offer interest.

Landscape Use This long-blooming, tough shrub is adaptable to landscape use, and in some cases has been over-used. Plant it in groups of three to five in foundation plantings or shrub borders, or use a single plant in mixed borders or rock gardens. It can be used as a low hedge. Its drought tolerance makes it a good choice for water-wise plantings. Plants are also salt tolerant and can be planted along sidewalks and roadways. It can also be used as large-scale groundcover on slopes and rocky soils.

Site Full sun to light shade in a wide range of soil types, but ideally moist and well drained. Withstands drought and high heat.

Culture Keep soil evenly moist until plants are established; after that, they can tolerate dry periods. A spring fertilizer application is beneficial. Shrubs require regular light pruning in early spring to look their best. No serious insect or disease problems. Spider mites can be a problem during hot, dry periods. Blast foliage with a garden hose to dislodge these pests.

Cultivars Many cultivars have been selected. Dakota Sunspot® ('Fargo') grows to only about 2 feet tall and wide and has golden-yellow flowers. 'Coronation Triumph' and 'Goldfinger' are com-

Potentilla fruticosa 'Fargo'

pact selections growing 2 to 3 feet tall and wide with bright yellow flowers. 'Jackmanii' is a vigorous, upright grower to 4 feet with deep yellow flowers. 'Snowbird' and 'Mount Everest' have white flowers. 'Tangerine' has orange flowers in shade or cool weather and yellow flowers in full sun and in summer heat. Zone 2.

Prunus americana
Wild plum
Zone 3

Native Habitat Moist to moderately dry soils along edges of woodlands mainly in the southern and western halves of Minnesota.

Size 15 to 20 feet tall, 15 to 20 feet wide

Description This single-trunked small tree has twiggy growth and suckers, which lead to large colonies in the wild. It has large, thornlike spur branches. The showy white flowers appear in 3- to 5-inch clusters April into early May, before the leaves are fully out, and fill the air with their perfume. The fruit is a large, fleshy, red plum, and it is edible. The dark green leaves turn golden yellow in fall.

Landscape Use The ideal spot for wild plum is a naturalized setting where you can allow it to colonize and enjoy the fragrant spring flowers and fall color. The fruits can be eaten fresh or used in jams, if you can get to them before the birds. With regular pruning of suckers and black-knot infections, it can be used as a specimen tree.

Site Adaptable to a wide range of conditions. Prefers a well-drained soil in full sun.

Culture Prune out root suckers if you don't want thickets to form. Black knot, a fungal disease, can be a problem on all *Prunus* species. It appears as a swollen portion of the stem tissue that matures into a hard, black area on the stem. Cut knots out about 4 inches below the knot and destroy the upper portion of the stem.

Other Species *P. nigra* (Canada plum) is native to rich, moist soils in scattered spots in all but southwestern Minnesota. It is rarely used as a landscape plant, but the cultivar 'Princess Kay' is an excellent small tree growing to only about 15 feet. The showy, double, white flowers, among the first blooms to appear in spring, are set off nicely by the dark bark. It originated near Grand Rapids, Minnesota. Zone 3.

P. pensylvanica (pin cherry) is native to dry soils and open hillsides throughout most of Minnesota. It grows up to 30 feet in height as a single-trunked tree with a narrow crown. The green cherries turn bright red at maturity and can be used for jams and jellies. Fall color is purplish red. It is a food source for wildlife. 'Stockton' is a hard-to-locate, double-flowered, round-headed form with bright red fall color. Zone 3.

P. pumila (sand cherry) is native to dry soils and open hillsides throughout most of Minnesota. This shrub grows 1 to 3 feet tall and is covered with white flowers in spring before it leafs out. The plump, roundish cherries turn a deep blackish purple. The astringent fruits can be used for jellies and jams and are a favorite wildlife food. Zone 3.

P. virginiana (chokecherry) is native on a wide variety of soils throughout Minnesota. It grows 15 to 35 feet tall and up to 25 feet wide as a thicket-forming shrub or small tree. White flowers appear in 2- to 3-inch racemes in May. Fruits are yellow to red cherries turning nearly black. They are relatively sweet when fully ripe and can be used to make wines, syrups, and jellies. Use chokecherry for naturalizing at wood edges or for bird food. 'Schubert' ['Canada Red'] has leaves that turn coppery red as they mature. It grows to only 20 feet tall and makes a good small tree in the lawn where the suckers are mown off regularly. In the garden, these suckers will need to be removed. Zone 3; trial in zone 2.

Prunus americana

Prunus nigra 'Princess Kay'

Rhus glabra
Smooth sumac
Zone 3

Native Habitat Dry or poor soils along forested edges throughout most of Minnesota.

Size 10 to 15 feet tall, 10 to 15 feet wide

Description Smooth sumac is a multiple-trunked deciduous shrub with an interesting branching habit. The dark green compound leaves are up to 24 inches long and have a tropical look. They turn a vivid orange, purple, yellow, and red in fall. Flowers are greenish in dense, upright panicles 4 to 8 inches long in July and August. On female plants, they turn into scarlet, cone-shaped clusters in September and October and persist into winter.

Landscape Use This is one of the first plants to turn color in fall, sometimes in mid-August, and this is its greatest asset. It spreads rapidly by suckers to form large colonies, and is good for stabilizing slopes to prevent erosion. Suckers are removed by regular mowing, so it can be grown next to lawns. It is tolerant of both salt and drought and can be used along roadsides. Animals and birds eat the fruits.

Site Almost any soil except constantly wet or boggy sites in full sun.

Culture Smooth sumac spreads quickly by underground roots that send up new trunks. Once established, it is hard to eradicate. Male and female flowers are on separate plants, so not all plants produce showy fruit clusters. Fall color will be poorer on rich sites, so avoid fertilization. Plants can be trained into interesting small trees. Choose a plant with a strong, straight leader and prune it annually to maintain shape. Suckers will need to be removed each spring to keep it from turning into a thicket.

Cultivars and Other Species
'Laciniata' is a female cut-leaf selection. 'Morden Select' is slower growing and only reaches 6 feet in height. Zone 3.

Rhus glabra 'Laciniata'

R. typhina (staghorn sumac) is native to dry or poor soils along forest edges throughout most of the eastern half of Minnesota. It can grow to 25 feet tall. It mainly differs from *R. glabra* in having dense, velvety hairs on its stems. Individual leaves are also narrower and more finely toothed. It has the same great fall color, usually more orange than red, and interesting fruits. It is the same plant as *R. hirta*. 'Laciniata' is a female selection with deeply cut leaves that colors later in the season. Zone 4; trial in zone 3.

Rosa blanda
Smooth wild rose
Zone 2

Native Habitat Dry to moist woods, prairies, and outcrops throughout Minnesota.

Size 2 to 6 feet tall, 4 to 6 feet wide

Description Smooth wild rose forms dense, impenetrable thickets in favorable situations. The stems are thornless or with scattered bristles toward the base of the plant. Flowers range from rosy pink to white and reach 2 to 3 inches across. They are fragrant and start blooming in June with some repeat bloom throughout the summer. The fruits, called "hips," resemble small apples and are about a half inch in diameter; they turn a nice red in fall and persist well into winter.

Landscape Use Plants make a nice addition to prairie or wildlife gardens. They can be used for naturalizing, massing, and in shrub borders, with some control of suckering. The hips are an important winter food source for wildlife.

Site Prefers rich soil in full sun to light shade, but will grow on poorer soils.

Culture Give plants a spring application of fertilizer. Suckers may need to be pruned. No pest problems.

Other Species *R. arkansana* (prairie rose) is native to dry to well-drained prairie soils in all but far north-central Minnesota. It is a spiny shrub growing about 2 feet tall and spreading 4 to 8 feet wide with stiff canes and dark green leaves. The 2-inch, solitary, soft pink flowers appear late spring to early summer and are slightly fragrant. They are followed by dark red hips. This rapidly spreading plant makes a good tall groundcover or soil stabilizer. Zone 3.

R. acicularis (prickly rose) is native to sandy soils in coniferous forests in northeastern Minnesota. It is a dense, prickly shrub growing about 3 feet tall with dark pink, fragrant flowers in June. It is somewhat shade tolerant and can be used as groundcover in slightly acidic soils high in organic matter. Zone 2.

Rosa blanda

Salix discolor
Pussy willow
Zone 3; trial in zone 2

Native Habitat Wet soils along shores, swamps, and wetlands throughout Minnesota, except for the far southwestern corner.

Size 10 to 20 feet tall, 10 to 15 feet wide

Description Pussy willow is a shrub or small tree with multiple trunks and an irregular crown. The narrow leaves are shiny green turning yellowish in fall. Its most ornamental feature is the 1-inch silvery gray catkins that appear before the leaves in early April. Twigs are reddish purple to dark brown.

Landscape Use Most people grow this shrub for its charming catkins, which are often part of spring floral arrangements. Although not as large their European cousins, they are still showy and a welcome sign of spring. Pussy willow is good for naturalistic plantings, in wet soils, or in mixed shrub borders, where it can blend in after flowering.

Site Thrives in moist to wet soils in full sun.

Culture This shrub requires little care and is pest free. Soil should be amended with peat moss or compost before planting to ensure adequate moisture retention. Mulch to maintain the necessary soil moisture. Plants can be cut back to the ground in early spring after enjoying the catkins. This keeps plants smaller and encourages long stems with showy catkins that are good for cutting.

Other Species *S. humilis* (prairie willow) is native to wet and moist soils in the same areas as *S. discolor*. The showy catkins are fat and fluffy with tiny bright yellow flowers, giving the entire shrub a yellow glow in early to mid-April. Use it for massing in low areas. Zone 2.

Salix humilis

Salix discolor

Sambucus canadensis
Common elder
Zone 2

Native Habitat Wide variety of soils mainly in the southern half of Minnesota.

Size 8 to 10 feet tall, 8 to 10 feet wide

Description This large shrub or small tree has multiple stems that are spreading or arching and a trunk that is usually short. The compound leaves have five to eleven leaflets, each 2 to 6 inches long. Small, white flowers are borne in dense, flat-topped clusters, up to 8 inches across, from June to July. The flowers are followed by small, berrylike, purple-black fruits borne in flat-topped clusters in late summer.

Landscape Use Even though the foliage, flowers, and fruits are attractive, common elder is difficult to utilize in most formal landscapes because of its unkempt growth habit. It suckers profusely and must be pruned regularly to keep it looking neat. Use it for naturalizing and screening. The fruits are used to make elderberry jam and are favorites of birds.

Site Does best on moist soils but it will tolerate dry soils, either acidic or alkaline.

Culture Drought tolerant once established, but plants will bloom and fruit better if given ample water and a spring application of fertilizer. It can be pruned back hard in spring to keep it in bounds. If you are growing elders for the fruit, you need to plant at least two different cultivars for cross-pollination.

Cultivars and Other Species
'Laciniata' is an attractive cut-leaf selection that does not fruit as well as the species. 'Adams' and 'York' were selected for their larger, more numerous fruits and are often planted by gardeners interested in the edible fruit. 'Aurea' has golden leaves that contrast nicely with the red fruits. It reaches 10 feet tall and wide, but is often pruned harshly every spring to force fresh foliage. 'Variegata' has narrow leaflets outlined in creamy white. It appreciates some protection from direct sun. Zone 3.

S. pubens (scarlet elder) is native to woodlands and woodland edges throughout most of Minnesota. This 10-foot-tall shrub is one of the earliest shrubs to leaf out and flower. Tiny greenish flowers and unfurling leaves appear in early to mid-April, noticeable

Sambucus canadensis 'Laciniata'

Sambucus pubens

because most other shrubs still have bare branches. The small, creamy white flowers are produced in erect pyramidal clusters in late April and early May and are followed by red fruits readily eaten by birds. It is shade tolerant and can be used for naturalizing. Zone 2.

Shepherdia argentea
Buffalo berry
Zone 2

Native Habitat Ravines and low prairies on sandy soils mainly along the western edge of Minnesota.

Size 15 to 18 feet tall, 15 to 18 feet wide

Description This loosely branched deciduous shrub has spiny branches covered with silvery scales when young and wedge-shaped leaves that are silvery on both surfaces. Showy red-orange fruits appear on female plants in August. The small, yellow-gold flowers in early to mid-April are not particularly showy (they are chunky and rough), but they do offer interesting texture to the early spring landscape.

Landscape Use Buffalo berry's tolerance of wind, drought, salt, and high pH soils makes it useful for tough sites. Use it for screening, naturalizing, and wildlife plantings. It can be used in shrub borders, where the silvery leaves offer a pleasant contrast. The fruits make good jelly, and the thorns provide excellent protection for nesting birds. It is a much better silver-leaved choice than the highly invasive and overplanted Russian olive (Elaeagnus angustifolia).

Site Dry or alkaline soils in full sun.

Culture Plants are dioecious, so plant both sexes to get colorful red fruits. Prune landscape plants in early spring; they can be cut to the ground. Does not need fertilizing. No serious insect or disease problems.

Shepherdia argentea

Spiraea alba
White meadowsweet
Zone 2

Spiraea alba

Native Habitat Moist to wet soils throughout Minnesota.

Size 2 to 5 feet tall, 2 to 5 feet wide

Description White meadowsweet has showy, white, branching clusters of numerous tiny flowers starting in June and continuing into September. They give the plants a fuzzy appearance. The twigs are dark brown and the leaves are blue green.

Landscape Use This moisture-loving shrub is a good choice for mixed borders or perennial gardens since it stays small and actually grows more like a perennial. It can be used along water edges and in bog gardens. The flowers are suitable for drying.

Site Prefers moist, fertile soils in full sun but will tolerate drier landscape conditions.

Culture Keep soil mulched to maintain necessary moisture. Plants will benefit from a spring application of fertilizer. White meadowsweet often dies back over winter and is best treated like an herbaceous perennial in garden settings. Cut plants to the ground in spring to keep them tidy and compact with good bloom, which occurs on new wood. This fast-growing plant will recover quickly from cutting back.

Other Species *S. tomentosa* var. *rosea* (steeplebush) is native to moist to moderately dry soils mainly in east-central Minnesota. It has light pink blooms in mid to late summer. Use it in the same landscape situations as *S. alba*. Zone 3.

Staphylea trifolia
Bladdernut
Zone 3

Staphylea trifolia

Native Habitat Moist, rich woodlands in southeastern Minnesota.

Size 10 to 15 feet tall, 8 to 10 feet wide

Description Bladdernut is an upright, heavily branched, suckering shrub or small tree with attractive striped bark. The leaves have three leaflets and are dark green in summer, turning pale yellow in fall. The interesting flowers appear in nodding panicles in mid-May, turning into light brown capsules in late summer.

Landscape Use Bladdernut makes a good understory shrub in woodland gardens. It can be used in shrub borders or trained as a small, vase-shaped tree. The 3-inch, inflated seedpods add interest in fall and winter and can be used in dried arrangements.

Site Prefers slightly acidic, moist, well-drained, fertile soils in sun or part shade but tolerates heavy shade, dry soils, and a higher pH.

Culture Bladdernut is an attractive shrub that adapts fairly easily to cultivation and should be used more often. Maintain soil acidity and moisture by mulching with a 2- to 4-inch layer of shredded pine bark or pine needles. Prune plants in early spring to shape or maintain small tree form. No serious insect or disease problems.

Symphoricarpos albus
Snowberry
Zone 2

Native Habitat Dry or rocky woodlands mainly in the northeastern half of Minnesota.

Size 3 to 6 feet tall, 3 to 6 feet wide

Description Snowberry's most ornamental feature is its berrylike, white, ½-inch fruits that appear September into November, persisting into winter. Branches often bend to the ground when loaded with fruits. It has bluish green leaves without any significant fall color. Flowers are small and pinkish and appear in June.

Landscape Use Snowberry's shade tolerance allows it to be planted along the edge of woods or under tall shade trees in wild gardens. The strong root system suckers freely and soon forms a thicket, making this shrub good for covering slopes and other tough-to-mow sites. The tangle of dense, twiggy stems has an unkempt look in winter, but these thickets provide excellent cover for wildlife, and the berries provide food for birds. Hummingbirds are attracted to the flowers. Fruiting stems are nice additions to autumn cut-flower arrangements.

Site Prefers a well-drained soil in part shade but it will grow in almost any light in any soil. Foliage is denser and the flowering heavier in full sun to partial shade.

Culture Common snowberry prefers consistently moist soil, but will survive in moderately dry conditions. Help maintain soil moisture by surrounding plants with organic mulch. Keep these suckering plants in check by cutting out entire stems or portions of stems in early spring to control the overall size and shape. A few leaf diseases, including anthracnose, blights, and powdery mildew, can cause spotting on the fruits and leaves but they rarely seriously harm plants.

Other Species *S. occidentalis* (wolfberry) is native to similar conditions. It grows about 3 feet tall and wide. It is seldom planted and difficult to obtain, but is excellent for attracting butterfly larvae. Zone 2.

Symphoricarpos albus

Symphoricarpos occidentalis

Viburnum trilobum
Highbush cranberry
Zone 2

Native Habitat Cool woods and thickets in all but southwestern Minnesota.
Size 10 to 12 feet tall, 10 to 12 feet wide
Description Highbush cranberry is a large, rounded shrub with gray, smooth branches. It has lovely white, lace-cap flowers up to 4 inches across late May into June. The indented dark green leaves turn a beautiful yellow orange to red in fall. The edible, deep red fruits appear in August, are showy, and often persist through winter.

Landscape Use Highbush cranberry has something to offer the landscape all year. Use it in shrub borders, as a specimen plant, in foundation plantings, and for screening. The edible fruits can be used for preserves and are attractive to birds.

Site Prefers a fertile, slightly acidic, well-drained soil in full sun to part shade but will tolerate less-than-ideal situations, even drought when mature. Shrubs will not flower and fruit as well in partial shade.

Culture This shrub readily adapts to landscape situations. Keep the soil evenly moist by mulching plants with wood chips or shredded bark. Plants will benefit from a spring application of fertilizer. Viburnums do not require a lot of pruning, but you can prune to shape or reduce height right after flowering. It is a good idea to remove a few of the older stems every three years or so to encourage new growth from the base.

Cultivars and Other Species
'Compactum' (also known as 'Bailey Compact') grows 6 feet tall and wide and is a good choice for hedging. 'Hahs' has a neat, rounded growth habit and good flower and fruit display. 'Wentworth' and Redwing™ ('J. N. Select') were selected for their heavy fruit production.

V. lentago (nannyberry) is native on a wide variety of soils along forest edges, on stream banks, and on hillsides throughout Minnesota. It is a 10- to 20-foot upright shrub or single-stemmed tree with drooping branches and a rounded crown. It grows well in full sun to deep shade, but fall color is best in sun. Leaves are shiny green and crinkled; fall color is reddish purple. Creamy white flowers appear in flat clusters 3 to 5 inches across in late May. The berrylike fruits start out yellow-green, turn dark purple in September and October, and persist into winter. They turn sweet after a frost and are eaten by birds. Nannyberry is easily pruned into a small tree, or it can be used in shrub borders or for screening. It will sucker and ramble, but not at an aggressive rate, and it is easy to prune. Plants may get powdery mildew, but it is purely an aesthetic problem.

It is a good choice where you want a shade-tolerant specimen tree. Unpruned specimens are good for naturalizing and use in woodland gardens. Zone 3.

V. rafinesquianum (downy arrow wood) is native to dry slopes and open woods in all but far southwestern Minnesota. It grows 6 feet tall and wide with white flowers and bluish black fruits. The leaves have an attractive reddish tint when young. Use it as an understory in woodland gardens, as an informal hedge, or in shrub borders. Zone 2.

Viburnum lentago

Viburnum trilobum

Viburnum rafinesquianum

Additional Native Deciduous Shrubs and Small Trees

***Alnus incana* ssp. *rugosa* (speckled alder)** is native to wet soils in all but the southwestern corner of Minnesota. It is a fast-growing small tree or shrub that grows to 15 to 25 feet tall with multiple thin trunks. It is shade tolerant and usually grows in dense thickets. Leaves are dull green, turning yellow in fall. The tiny, conelike seeds, which appear only after plants are eight to ten years old and then only every four years after that, are often used for jewelry or in artwork. Use it for naturalizing on wet sites. Zone 2.

***Chamaedaphne calyculata* (leather leaf)** is native to peaty soils in the northeastern half of Minnesota. It is an evergreen, rhizomatous semi-shrub growing 1 to 4 feet tall. It has white flowers in April and May. Plants require an acidic peaty or sandy soil that is constantly moist, such as in bog gardens. Zone 2.

***Epigaea repens* (trailing arbutus)** is native to acidic, sandy soils in northeast-ern Minnesota. It is a creeping shrub growing less than a foot tall with bright green leaves and fragrant, pinkish flowers in April and May. It is difficult to grow but can be tried as groundcover in woodland gardens with acidic soils. Zone 2.

***Kalmia polifolia* (bog laurel)** is native to acidic peat bogs in northeastern Minnesota. It grows 1½ to 2 feet tall and wide with narrow, dark green, evergreen leaves. Rosy red to pink flowers appear in late spring. It requires acidic, boglike conditions, which are difficult to reproduce in landscapes. Consider it for groundcover if you can provide the moist, acidic conditions it requires. Zone 2.

***Ledum groenlandicum* (Labrador tea)** is native to acidic bogs in the northeastern half of Minnesota. It grows to 3 feet tall and wide and has elongated, evergreen leaves, wooly underneath with brownish hairs. White flowers appear in small, flat-topped clusters in May. It can be used if you have a natural low area with acidic soil or an artificial bog. It needs a soil pH of 4.5 and constant moisture. Zone 3.

***Rubus parviflorus* (thimbleberry)** is native to moist soil on wooded hillsides and along stream banks in far northeastern Minnesota. It is a scrambling shrub 2 to 8 feet tall with attractive, fragrant white flowers and colorful orange to maroon fall foliage. Fruits are red berries that fall to the ground when ripe. It is moderately shade tolerant and can be used as groundcover on most soils. Zone 2.

***Vaccinium angustifolium* (low-bush blueberry)** is native to acidic, sandy soils or bogs in the northeastern half of Minnesota. It grows 6 to 12 inches tall. Flowers are white, summer foliage is glossy green, and fall color is an excellent red. It can be used as groundcover on acidic soils or grown for the small edible fruits. Zone 2.

Evergreen Conifers

When you have the potential for five or more months of snow cover each year, it is important to include landscape plants with winter interest. Evergreens excel in this area, but they are also valuable for the many other benefits they bring to a landscape. Their evergreen foliage provides a dark backdrop for flowering and fruiting herbaceous plants. They provide year-round screening, and many make good hedges. Many can be used in foundation plantings, where they offer a certain stability, and the shrub types offer interest and texture in shrub borders. Most provide shelter for birds and other forms of wildlife, and some provide food as well. The aromatic cut foliage is often used for indoor holiday arrangements.

Most people would agree that mature evergreens are a real asset when it comes to property value. However, their value quickly goes down if they are diseased or improperly pruned, which is often the case. It is so easy to fall in love with cute little "button" or pyramidal evergreens, but keep in mind than many landscape plants tend to outgrow their sites. They grow into large trees or spreading shrubs that aren't easy to keep small with pruning, and they are more susceptible to insect and disease problems when they are growing in a contained space. Most evergreens look best when they are allowed to grow naturally, without a lot of shaping. If you want an evergreen to stay within a certain size range, look for a dwarf or compact cultivar. This is much more effective than trying to restrict the growth of full-sized species.

When choosing an evergreen, consider hardiness, light requirements, rate of growth, attraction to wildlife, and winter appearance. Select a tree or shrub based on your soil conditions if you don't want to drastically change them. Keep in mind also that some evergreens are favorite food of deer.

Planting

You will have the best success planting northern-grown nursery plants that have been properly root pruned. They will survive transplanting best and start growing quickly. Bare-root trees must be planted in spring as soon as the soil can be worked. Balled-and-burlapped plants, container plants, and tree-spade trees can be planted anytime but the hottest days of summer, from mid-June to mid-August. Spring is still the best time to plant, however. If you have the opportunity to move a native evergreen, stick with one with a trunk less than 2 inches in diameter and move it in early spring.

Before planting, amend the soil with a good amount of organic matter such as compost, peat moss, or well-rotted manure. Mix this thoroughly with the planting-hole soil. Place the evergreen at the same depth it was growing in the container or burlap wrap. Bare-root plants should be planted so that the crown is level with the ground level. Newly planted evergreens should not need additional fertilizer, but it is a good idea to surround them with a 2- to 4-inch layer of organic mulch such as wood chips, shredded bark, or pine needles. Replenish the mulch as needed throughout the growing season.

Care

The first two or three years after planting, make sure the soil is evenly moist from spring until the ground freezes in fall. Once established, many evergreens can tolerate some dry periods, but don't hesitate to water as needed, especially in sandy soils. Always saturate the soil thoroughly with each watering to encourage deep rooting. To avoid browning of needles in winter, make sure the plants have plenty of moisture right up until the ground freezes.

Most young evergreens will benefit from a spring application of an acidic fertilizer. Spread a layer of rotted manure or compost around each plant or use a fertilizer such as cottonseed meal, Milorganite, or fish emulsion. If possible, allow needles to fall and decay under plants to return nutrients and acidity to the soil. Keep weeds pulled or smother them with organic mulch. Never cover the roots of these forest trees with plastic or rock.

Maintenance needs of evergreens differ depending on the species. Most shrubs will benefit from regular trimming to help maintain their natural shape. A light trimming is okay spring into early summer, but do not cut branches too far back. Stay in young green growth. Most evergreens do not generate new growth in older sections. Remove any dead, damaged, or diseased parts of evergreens at any time of year.

This planting of *Pinus resinosa* (red pine) provides winter interest as well and a windbreak.

Possible Problems

Native evergreens are susceptible to insects and diseases, some of which can become serious. The best defense against insect and disease problems is a vigorously growing plant. By choosing a tree or shrub appropriate for your site conditions and providing it with the necessary nutrients and ample water, most pest problems will not become serious in landscape situations. Many insect and disease problems are purely cosmetic and don't really threaten the life of the tree. On large trees, control measures are rarely practical. On young plants, keep an eye out for aphid feeding (spray foliage daily with a strong blast of the hose), iron chlorosis (acidify soil to lower pH), and scale (spray with a dormant oil in early spring).

Winter burn, or browning of needles, is common on evergreens and caused by insufficient water in winter. This can be due to a sudden drop in temperature, direct sun, desiccating winds, or drought conditions in fall. The sun, reflected from a white snow surface on a still day in February, may cause the temperatures in the leaves to rise as much as 50 to 60 degrees above air temperatures. If the sun goes behind a cloud or building, the temperatures drop suddenly and tissue within leaves is killed. Plant sensitive evergreens such as arborvitaes, Canada yew, and some junipers on sites where they will receive some winter shade, and make sure plants receive ample water going into winter.

Abies balsamea
Balsam fir
Zone 2

Native Habitat Moist soils, shaded forests, and along bogs in the northern half of Minnesota and in a few scattered spots in the southeastern corner.

Size 30 to 50 tall, 25 feet wide

Description Balsam fir is an aromatic and handsome tree with a continuous straight-tapering trunk from bottom to top. The spreading branches form a symmetrical, slender pyramid. Bark is smooth, grayish, and prominently marked by blisters filled with resin. Needles are flat with rounded points, dark green and lustrous above and silvery white beneath. Showy female cones are upright and bluish purple, 2 to 3 inches tall, and clustered near the tops of trees. Balsam fir is often found growing with white spruce in native stands.

Landscape Use Balsam fir is especially attractive as a young tree, which is good because it is one of the slower growing evergreens (6 inches or less a year). It tends to be a little sparse looking as it ages, especially when you compare it to a spruce. It is a good choice for naturalizing on a north slope or in moist conditions. Plant a group of three in a large shade garden. Trees provide winter cover for wildlife. It is one of the most popular Christmas trees, since it holds its needles well and smells wonderful.

Site Does best in cool, damp places on moisture-retentive, acidic soils in part shade to sun. Keep it away from hot, drying winds.

Culture Amend soil with organic matter before planting and acidify if necessary. Landscape trees should be mulched and watered as needed. Older trees will shade and mulch their own root systems. Balsam fir does not have any serious insect or disease problems and rarely needs pruning. Winter browning is rarely a problem if plants are sited right.

Cultivars 'Nana' is a mounded, low-spreading form growing only 18 to 24 inches tall that is available from some specialty nurseries. Zone 3.

Abies balsamea 'Nana'

Abies balsamea

Juniperus communis var. depressa
Bush juniper
Zone 2

Native Habitat Sandy soils on steep bluffs, river bottoms, and swamp margins in all but southwestern Minnesota.

Size 1 to 3 feet tall, 3 to 10 feet wide

Description This spreading shrub has a sharp, angular form and a rather coarse texture. The dense, awl-shaped needles are green in summer and take on a purplish color in fall. Bark is reddish brown, peeling off in strips. Plants are dioecious, with females having the potential to produce showy berries.

Landscape Use Junipers are among the best evergreen shrubs available for northern landscapes. Use the coarser species for naturalizing or as groundcover. The cultivars are better choices for use in shrub borders, foundation plantings, and rock gardens. All types provide winter cover for wildlife. Avoid planting junipers under drip lines, where crashing snow can damage them. Native junipers tolerate road salt better than most evergreens.

Site Prefers neutral to slightly acidic, well-drained soil in full sun. Tolerates drought and wind well. Foliage will be sparser in shade.

Culture Make sure plants are well watered going into winter to reduce chances of winter burn. Plants can easily outgrow their space. Regular pruning of the branch tips in June will keep them from becoming overgrown and needleless in the middle. It does not work well to try to prune back overgrown plants. Browned needles may need to be combed out from plants to keep them looking neat. Trim upright forms so that the upper part is slightly narrower than the base to allow light to get to the lower foliage. Bagworms can be a problem: remove and destroy any bags as soon as you see them. Deer are usually not interested in junipers.

Cultivars and Other Species
'AmiDak' (Blueberry Delight™) was selected for its ability to produce prolific blue fruits when pollinated. Zone 3. 'Depressa Aurea' has new growth with a golden-yellow color. Zone 4. 'Repanda' is a compact, rounded form to 1 foot tall, spreading 6 feet wide. Zone 4.

J. horizontalis (creeping juniper) is native to sandy soils widely scattered in Minnesota. This low, creeping shrub is usually less than 1 foot tall, spreading 3 to 6 feet to form dense ground-cover. Leaf color is bluish green to steel blue, taking on purplish tints in fall. Juniper blight is a disease that can show up during wet springs, turning portions of the plant brownish. Snip off diseased parts and dispose of them. Some cultivars are more resistant to blight. Creeping juniper is on the state's special concern list, so do not dig plants from the wild. Zone 2. Many cultivars have been selected for wonderful textures and foliage color and a variety of shapes and forms. They are good landscape plants that can be used as groundcovers on sandy soil, in foundation plantings, and in rock gardens. 'Bar Harbor' spreads up to 10 feet wide with trailing bluish green branches that turn purple in fall. 'Blue Chip' is low growing with good blue color throughout the year. 'Hughes' is somewhat resistant to juniper blight. It has distinct radial branching and silvery blue leaves and spreads up to 10 feet. 'Wiltonii' is a low form (3 inches tall by 4 feet wide) with intense silvery blue color. All cultivars are hardy into zone 3.

J. virginiana (eastern red cedar) is native to dry soils and open hillsides in the southern half of Minnesota, often on river bluffs where few other trees are found. It grows 25 to 40 feet tall with a straight trunk and a broad, conical head. The thin, reddish brown bark peels off in long strips. Leaves are dark green and scalelike; newer growth may be sharp-pointed and whitened underneath. Dark blue berrylike cones on female trees are a favorite winter food of wildlife, and trees provide winter shelter for birds. Eastern red cedar is the alternate host of cedar apple rust, a disease that causes strange-looking orange galls on trees. Snip off any plant parts infected with the galls and destroy them. Use this durable tree for screening, windbreaks, or in natural settings on tough sites. Zone 3. 'Canaertii' stays at 20 feet in height with a spread of 8 feet. It has a dense, pyramidal form and dark green leaves. 'Gray Owl' is a spreading form with silvery gray foliage. It grows 4 feet tall and 5 feet wide. Both cultivars are hardy into zone 4.

Juniperus communis var. depressa 'AmiDak'

Juniperus horizontalis 'Blue Chip'

Juniperus virginiana

Picea glauca
White spruce
Zone 2

Native Habitat A variety of soils, often on banks of lakes and streams, sometimes in pure stands in the northern half of Minnesota with a few southeastern locations.

Size 40 to 60 feet tall, 40 feet wide

Description This large tree is pyramidal when young, becoming narrower as it matures. Bark is dark gray or grayish brown and scaly. The stiff, bluish green needles are crowded along branchlets. Slender, 2-inch cones hang from branches.

Landscape Use Spruces are vital to northern landscapes, and are long-lived if properly cared for. Their large size restricts their use, but they can be used for screening, windbreak and shelterbelt plantings, and background plantings. Don't plant a spruce tree if you want to mow right up to the trunk. Their natural shape is pyramidal with branches all the way to the ground, not palm-tree-like. They provide food and cover for wildlife and are good Christmas trees. Dwarf and compact forms can be used as specimens or accents in the landscape and in rock gardens.

Site Does best on well-drained, slightly acidic soils in full sun. Foliage is thinner in shade. Keep away from winter winds and road salt.

Culture Give white spruce plenty of space. Crowded trees will drop needles. Keep soil evenly moist. Trees do poorly in overly wet soils and in drought conditions. Spruces require little or no pruning. If you want to shape a spruce tree, do a little trimming in June just as the new growth is hardening, rather than a lot all in one year. If possible, keep snow off branches in winter to reduce branch breakage. Sawflies can leave bare spots on trees. Remove them by hand as soon as you see them. Cankers show up as wounds on trees and branches and lead to defoliation. Prevent cankers by growing healthy trees. Once a tree is infected, there is little you can do.

Cultivated Varieties and Other Species 'Conica' is a compact, slow-growing form that grows about 5 feet tall and 3 feet wide. Protect it from winter sun and winds, or shade it with burlap to reduce winter burn. Zone 4. *P. glauca* var. *densata* (Black Hills spruce) seldom grows taller than 40 feet and is denser and more drought tolerant. Foliage is a little bluer. It is a good choice for landscape use. It is often sold as 'Densata'. Zone 3.

P. mariana (black spruce) is native to wet soils in swamps and bogs in the northeastern half of Minnesota. It usually stays at 20 to 30 feet, but can grow over 50 feet in favorable conditions. It has a narrow, pyramidal shape and widely spaced branches that are somewhat drooping with bluish green needles. It is fairly drought tolerant in landscape situations if well mulched but does best on wetter soils. Zone 2. 'Nana' is a slow-growing, mounded cultivar growing 2 feet tall and 3 feet wide. It needs partial shade. Zone 3. 'Ericoides' is a slow-growing dwarf that eventually becomes rounded and flat-topped. Zone 3.

Picea glauca var. *densata*

Picea mariana 'Ericoides'

Pinus strobus
White pine
Zone 3

Native Habitat Wide variety of soils from dry and sandy to moist upland sites mainly in the eastern half of Minnesota.

Size 60 to 100 feet tall, 50 feet wide

Description White pine is the largest conifer in Minnesota. It has a pyramidal shape with whorls of horizontal branches evenly spaced along the trunk. The bark is thin, smooth, and greenish gray on young trees, but thick, deeply furrowed, and grayish brown on older trees. The soft, flexible, gray-green needles are 2½ to 5 inches long and occur in clusters of five. Cones are 4 to 8 inches long, thick, and usually gummy.

Landscape Use There's nothing like the scent of pine trees or the sound of wind whistling through their branches. They can be used as specimens on large landscapes, but are usually used for screening and windbreaks. They don't adapt well to urban conditions.

Site Best growth is on fertile, acidic, well-drained soils in full sun. Pines do not need protection from winds, but keep them away from road salt.

Culture Pruning is not recommended for white pine, but it does help form more-compact growth on cultivars. Prune in spring when the candles are beginning to lengthen, pinching or cutting back one-third to one-half of the candles, no more. Although native trees are susceptible to several insect and disease problems, landscape trees are usually not bothered. White pine blister rust shows up as powdery red rust on the bark. The alternate host is *Ribes* species (gooseberries and currants). Don't plant these where white pines are a priority. Pick off sawflies as soon as you see them. Scale insect damage shows up as whitish flecks on needles. For serious infestations, you may need to spray plants with dormant oil or lime sulfur in late fall or early spring. Deer browsing can be a problem, and young trees can get winter burn on exposed sites. Needles in the center of trees brown before falling, and this is often confusing to uneducated homeowners who think their tree is dying.

Cultivars and Other Species 'Blue Shag', 'Compacta', and 'Nana' are rounded, dense shrubs growing 6 to 8 feet tall and 6 feet wide; 'Compacta' has softer foliage. 'Fastigiata' is a columnar form growing about 25 feet tall and 8 feet wide. 'Pendula' is a 10-foot weeping form that must be staked when young. It takes several years to really look nice. All cultivars are hardy in zone 4.

P. resinosa (red pine) is native to dry, sandy soils, often in pure stands, mainly in northeastern Minnesota. Mature trees are 40 to 80 feet tall or more and have an open, rounded, picturesque head. Bark develops reddish brown plates as it matures, giving this tree its common name. The stiff, 4- to 6-inch needles appear in clusters of two. Cones are about 2 inches long, light brown fading to gray and free of resin. Red pine is the official Minnesota state tree and is impressive when seen in windbreaks and shelterbelts. It thrives on sandy loam and dry soils in full sun and is disease and insect resistant. Winter burn can be a problem on younger trees. Zone 3. 'Wissota' is a dwarf form introduced by the University of Minnesota that grows about 6 feet tall and wide. Use it as a specimen or in rock gardens. Zone 3.

P. banksiana (jack pine) is native to acidic, sandy soils of low fertility in the northeastern half of Minnesota. It grows to 60 feet, but usually stays closer to 20 to 40 feet in landscape situations. It is an open tree with many small dead branches that often remain on trees for many years. Bark is a dull red-brown to black color with loose scales or plates. Needles are flat, grayish green, twisted two in a bundle. Cones are strongly curved, brown when ripe, turning gray. It is tolerant of poor sandy soil and shade, thriving in sites unsuitable for white or red pine. It is fast growing when young, making it good for windbreaks and screening. Its stark, open growth habit can be adapted to landscape use, especially with modern architecture. Zone 2. 'Uncle Fogy' is an interesting University of Minnesota introduction with an irregular, often weeping, form. It grows about 6 feet tall and 10 feet wide. Use it as a specimen or accent plant. Zone 3.

Pinus strobus

Pinus banksiana 'Uncle Fogy'

Taxus canadensis
Canada yew
Zone 2

Taxus canadensis

Native Habitat Cool, rich, damp woods and bog margins in the northeastern half of Minnesota.

Size 3 to 6 feet tall, 6 to 8 feet wide

Description This low-spreading, open, irregular shrub has attractive dark green needles that usually take on a purplish cast in fall. Plants are dioecious, with female plants producing red berrylike fruits.

Landscape Use Canada yew is good for naturalizing under large trees, especially pines. It can be planted along the north or east side of buildings in informal foundation plantings, in shrub borders, or woodland gardens. Plants are slow growing and easy to keep under control with pruning.

Site Prefers a slightly acidic soil in full to partial shade out of winter winds.

Culture Although Canada yew is hardy, it is subject to winter burn if planted in the wrong conditions. Make sure plants have winter shade or reliable snow cover. If you can't provide these conditions, protect plants with a burlap screen in winter. Landscape plants should be well mulched. Unlike other evergreens, yews have the ability to sprout new growth from fairly old wood, so they can be cut back severely and still develop good foliage. It is still better to prune plants regularly rather than resort to drastic cutting back. The foliage and seeds of yews are poisonous and should be kept from small children. Unfortunately, this does not deter deer. This plant is difficult to find in the nursery trade.

Thuja occidentalis
White cedar, arborvitae
Zone 2

Native Habitat Moist or wet soils often in pure stands in the northeastern quarter of Minnesota.

Size 30 to 50 feet tall, 30 feet wide

Description This upright, pyramidal evergreen has dense, scalelike, green to yellowish green foliage arranged in flat, fanlike branches. On mature trees the bark is gray to reddish brown, separating in long shreddy strips, and the trunk is often twisted. Foliage and bark are aromatic. Foliage often looks slightly yellow, purple, or brown in winter but returns to green in summer.

Landscape Use The large size of the species limits its use to screening and windbreaks. The cultivars are good choices for hedging, specimen plants, and foundation plantings. All provide excellent shelter for birds.

Site Does best in a moisture-retentive soil in full sun or partial shade. In deep shade, plants are open with sparse foliage. Keep plants away from dry, windswept locations.

Culture Arborvitaes are easy to grow in any moisture-retentive soil. They are slow growing and long-lived. They are subject to winter burn, especially when planted on the south side of buildings,

This planting includes, pictured left to right, *Thuja occidentalis* 'Holmstrup', 'Techny', 'Hetz Midget', and 'Woodwardii'.

but it is easy to prune out in spring since it occurs on outside leaves. Do not confuse winter burn with the normal browning of inside leaves as plants age or the normal color change arborvitaes go through in winter. Prune just after new growth has emerged. Formal hedges can be pruned again once later in the season, but do not prune in late fall. Plants are susceptible to damage from heavy snow or ice. Try to plant them away from the roofline. Deer are a serious problem, often completely defoliating the lower parts of trees.

Cultivars Many cultivars have been selected for foliage color and growth habit. Here are some good choices for northern use: 'Hetz Midget' is a dense, globe-shaped selection that grows 2 feet tall and 3 feet wide. 'Holmstrup' is a tough, upright grower that stays under 8 feet tall. 'Techny' is a broad, pyramidal tree 25 feet tall with good dark green color; a good choice for a tall hedge. 'Wintergreen' and 'Wareana' grow to 15 feet tall and are good for mid-height hedging. 'Woodwardii' is a true globe, growing 5 to 6 feet tall and wide. All cultivars hardy in zone 3.

Tsuga canadensis
Eastern hemlock
Zone 3

Native Habitat Cool, acidic, moist soils containing considerable organic matter in northeastern Minnesota.

Size 40 to 60 feet tall, 30 to 50 feet wide

Description Eastern hemlock is a willowy, flexible evergreen with horizontal branches that droop gracefully. The flattened, deep green sprays of soft, short needles hold their color well all year. The cinnamon-red bark is deeply divided into narrow ridges. The flat, blunt needles are borne in many rows and are dark yellow green above, lighter colored below. The ½- to ¾-inch-long cones hang down from the ends of twigs.

Landscape Use Eastern hemlock is a long-lived, fine-textured evergreen that has much to offer if it has the proper growing conditions. Use it as a specimen or in odd-numbered groupings. It can be used as a hedge or a background or screening plant.

Site The best site is partially shaded with a cool, moist, well-drained, acidic soil. Eastern hemlock can be grown in full sun as long as it has a highly organic soil and no strong, drying winds. Avoid windswept sites and polluted conditions.

Culture Eastern hemlock is one of the few evergreens that can tolerate shade, but it is sensitive to environmental extremes. Avoid heat, drought, wind, road salt, air pollution, and poor drainage. It tolerates winter cold but can be damaged by unseasonable frosts. Sun scorch can occur when temperatures reach 95 degrees, killing ends of branches. Pruning is not usually necessary, but trees are amendable to late-spring pruning. It is on Minnesota's special concern list so do not dig plants from the wild. Purchase plants propagated from a local seed source for best results.

Tsuga canadensis

Tsuga canadensis

Vines

Vines serve both ornamental and functional roles in the landscape. They are a good way to bring the dimension of height to areas not suitable for trees, such as in tight courtyards or next to buildings. Their verdant covering softens arbors and trellises, often providing shade for areas underneath. They also serve as screening, covering unsightly landscape elements such as chain-link fences and compost and storage structures. There are only a few native vines suitable for landscape use, and most are rampant growers without showy flowers, best suited to covering unsightly objects.

To use vines correctly, you must understand their climbing characteristics. Some, like wild grape, have tendrils that attach the vines to supports. Other vines climb by twining or wrapping themselves around supports; bittersweet is an example. Woodbine sends out sucker-like disks that attach to supports. The support structure should be in place before planting. If you install a fence or trellis after the vine is planted, you run the risk of damaging the roots.

When selecting a vine, consider hardiness, light requirements, growth rate, and seasonal aspects like flowers, fruits, fall color, attraction to wildlife, and winter appearance.

Planting

Most vines should be planted in early spring. If you have the opportunity to move a native plant, move it in early spring. Bare-root plants must be planted in spring as soon as the soil can be worked. Container plants can be planted anytime but the hottest days of summer, during July and August. Spring is still the best time for planting, however.

Before planting, amend the soil with a good amount of organic matter such as compost, peat moss, or well-rotted manure. Mix this thoroughly with the planting-hole soil. Place the vine at the same depth it was growing in the container. Bare-root plants should be planted so that the crown is level with the ground level. Newly planted vines should not need additional fertilizer. It is a good idea to surround all newly planted woody plants with a ring of organic mulch 2 to 4 inches thick. Good mulches are wood chips, shredded bark, and pine needles. Replenish the mulch as needed throughout the growing season. Once the vines are tall enough, they will probably need to be trained to their supports. You can tie them with soft twine or gently wrap the young stems around or through supports.

Care

The first two or three years after planting, make sure the soil is evenly moist from spring until the ground freezes in fall. Once established, most native vines can tolerate some dry periods, but don't hesitate to water as needed, especially in sandy soils. Always saturate the soil thoroughly with each watering. Most vines are rampant growers that need little or no fertilizer once established, but young plants will benefit from a spring application of fertilizer. Spread a layer of rotted manure or compost around each plant or use Milorganite or fish emulsion. Keep weeds pulled or smother them with organic mulch.

Most vines will benefit from some selective pruning while they are young to develop a good shape. The best time to prune is early spring. Remove any dead, damaged, or diseased parts at any time of year, cutting back to just above a healthy, outfacing bud. Vines that have become overgrown can be renewal pruned by cutting plants back to about 4 inches from the ground in early spring. Surround them with a layer of compost or rotted manure and water them well to help them recover.

Possible Problems

Few insects or diseases bother vines. Powdery mildew can affect some vines, especially when the dense foliage remains wet for long periods. Snip out the infected areas and dispose of the foliage. Thin out plants to allow for better air circulation and drying of foliage. Leaf spots may leave plants somewhat defoliated by late summer, but this is rarely a serious problem. Aphids often feed on young leaves in hot, dry weather. Dislodge them by spraying the foliage daily during hot weather with a hard spray from the hose.

Native vines may not always be high on the list for their ornamental value, but they can be functional in the landscape, often used to cover chain-link fencing and other undesirable features.

Celastrus scandens
American bittersweet
Zone 3

Native Habitat Moist clearings and thickets throughout most of Minnesota.

Height Up to 25 feet

Description This vigorous climbing vine's most ornamental feature is the showy fruits, which appear on female plants in fall. The fruit has an outer yellow-orange husk that opens to expose the red aril. Outer husks remain attached to the base of fruits, adding contrast. The fruits hang in 2- to 3-inch-long clusters like small bunches of grapes. American bittersweet has nondescript greenish white flowers in late spring and wedge-shaped, green summer foliage.

Landscape Use Bittersweet is mainly grown for its showy fall fruits, but it can be used as a screen in difficult sites, to cover a steep bank, or to scramble over stone walls or fences. It is a strong plant that can overtake less-vigorous plants growing nearby. The twining, fast-growing shoots will coil around each other if there is nothing else available. Songbirds eat the fruits.

Site Adaptable, but prefers moist, well-drained soils. Fruiting is best in full sun, but plants will grow in light shade.

Culture Since plants are dioecious, both male and female forms must be planted to ensure fruit production. It needs a strong trellis or fence to support the prolific growth. Prune this vigorous vine back a bit each spring to allow good sun exposure to all the stems. Harvest branches as soon as they reach maturity for use indoors. American bittersweet has become invasive in warmer areas of the country, but is pretty well behaved in zone 4 and colder regions. Be sure you are planting the native species and not the more aggressive introduced *C. orbiculatus* (Oriental bittersweet); they look similar.

Celastrus scandens

Clematis virginiana
Virgin's bower
Zone 3

Native Habitat Open or semi-shady woods and thickets throughout most of Minnesota.

Height Up to 20 feet

Description This vigorous vine has compound leaves with three leaflets. The small but numerous creamy white flowers appear in dense panicles in late June and July. The flowers turn to ornamental, fluffy seed clusters in August and September that persist into winter.

Landscape Use Virgin's bower is one of the best native vines for landscape use. It can be grown as a traditional vine, growing up a trellis or fence, planted at the base of mailbox posts, or used as groundcover on steep banks or other tough sites. It is a prolific grower that will quickly cover cut stumps or other items that need screening. It will grow up into nearby trees and shrubs, and is sometimes used this way to bring color to spring-blooming plants that are finished blooming.

Site Adaptable. Prefers well-drained soil in full sun to light shade.

Culture Virgin's bower climbs by twining its stems around its support. Without a support, it will creep along the ground more like a groundcover.

Other Species *C. occidentalis* var. *occidentalis* (western blue virgin's bower) is native to acidic pine forests in far northeastern and southeastern Minnesota. It can grow up to 20 feet long with blue to violet, nodding, 1½-inch flowers late spring to early summer. It is a nice plant for a shady trellis or along the ground in woodland gardens if you can provide it with the shady, acidic soil conditions it requires. It is difficult to find retail sources. Zone 3.

Clematis virginiana

Parthenocissus quinquefolia
Woodbine, Virginia creeper
Zone 2

Parthenocissus quinquefolia

Native Habitat Edges of woods widely scattered in the southeastern quarter of Minnesota.

Height Up to 30 feet

Description Woodbine has large, palmately compound leaves with five leaflets that turn brilliant red early in fall. Small white flowers appear in inconspicuous panicles in July and August. The bluish black berries are showy after the leaves have fallen. The woody stems are quite large on older plants.

Landscape Use This vine is too aggressive for many landscape situations but is good for covering in less-formal situations. Use it for screening fences and other unsightly items. It is the only native vine that will adhere to flat surfaces such as brick walls. It is not a good choice for wood-sided buildings, however, since it can stain the wood. Choose a spot where the showy fall color will stand out. It will grow as groundcover and can be used for screening on tough sites. Birds eat the fruits, and seedlings can appear throughout the landscape.

Site Will grow on almost any soil in sun to shade.

Culture Woodbine is a fast-growing vine that clings by tendrils that end in sucker-like disks. The tendrils can twine around wire or twigs, and the disks adhere to bark or walls, so it will grow up them without a support. As it becomes established, aerial rootlets grow out of stems to hold more permanently. It can be cut back to the ground in spring to keep it from developing a thick, woody trunk and growing out of bounds.

Cultivars 'Engelmannii' and 'Saint-Paulii' have smaller leaflets. 'Saint-Paulii' is best for a brick or stone wall because its sucker-like disks are better developed. 'Variegata' has green-speckled leaves but they often revert to green. Cultivars hardy in zone 3.

Vitis riparia
Wild grape
Zone 2

Vitis riparia

Native Habitat Riverbanks and rich woods throughout most of Minnesota.

Height Up to 30 feet

Description This vigorous vine has traditional grape leaves 3 to 7 inches long that turn an attractive yellow-gold color in fall. Fragrant flowers appear in panicles in May and June and are followed by the showy, purple-black clusters of fruits. Individual grapes are less than a half-inch in diameter and have a whitish bloom. Stems of mature plants can be 2 inches or more in diameter and have attractive shredded bark.

Landscape Use Wild grape is one of the more attractive native vines. It can be used for screening or covering a large, sturdy pergola. It needs a substantial structure to support the woody stems. The small fruits can be used for jelly and are a favorite of many species of wildlife.

Site Grows well in most soils in full sun to light shade. Fruiting is best in full sun.

Culture The stems of wild grape attach by twining and tendrils. Provide a strong support for the woody stems. Keep stems thinned out to reduce chances of powdery mildew.

Bibliography

Books

Art, Henry W. *The Wildflower Gardener's Guide: Midwest, Great Plains, and Canadian Prairies Edition.* North Adams, MA: Storey Communications, 1991.

Burrell, C. Colston. *A Gardener's Encyclopedia of Wild Flowers.* Emmaus, PA: Rodale Press, 1997.

Coffin, Barbara, and Lee Pfannmuller, editors. *Minnesota's Endangered Flora and Fauna.* Minneapolis: University of Minnesota Press, 1988.

Cox, Jeff. *Landscaping with Nature: Using Nature's Designs to Plan Your Yard.* Emmaus, PA: Rodale Press, 1991.

Cullina, William. *Native Trees, Shrubs, and Vines: A Guide to Using, Growing, and Propagating North American Woody Plants.* Boston: Houghton Mifflin, 2002.

Cullina, William. *The New England Wild Flower Society Guide to Growing and Propagating Wildflowers of the United States and Canada.* Boston: Houghton Mifflin, 2000.

Dirr, Michael A. *Manual of Woody Landscape Plants: Their Identification, Ornamental Characteristics, Culture, Propagation, and Uses.* Fifth edition. Champaign, IL: Stipes Publishing, 1998.

Imes, Rick. *Wildflowers: How to Identify Flowers in the Wild and How to Grow Them in Your Garden.* Emmaus, PA: Rodale Press, 1992.

Johnson, Lorraine. *Grow Wild! Low-Maintenance, Sure-Success, Distinctive Gardening with Native Plants.* Golden, CO: Fulcrum Publishing, 1998.

Jones, Samuel B., Jr., and Leonard E. Foote. *Gardening with Native Wild Flowers.* Portland, OR: Timber Press, 1990.

Moyle, John B., and Evelyn W. Moyle. *Northland Wildflowers: The Comprehensive Guide to the Minnesota Region.* Minneapolis: University of Minnesota Press, 2001.

Ownbey, Gerald B., and Thomas Morley. *Vascular Plants of Minnesota: A Checklist and Atlas.* Minneapolis: University of Minnesota Press, 1991.

Peterson, Roger Tory, and Margaret McKenny. *A Field Guide to Wildflowers: Northeastern and North-central North America.* Boston: Houghton Mifflin, 1968.

Rose, Nancy, Don Selinger, and John Whitman. *Growing Shrubs and Small Trees in Cold Climates.* New York: Contemporary Books, 2001.

Shirley, Shirley. *Restoring the Tallgrass Prairie: An Illustrated Manual for Iowa and the Upper Midwest.* Iowa City, Iowa: University of Iowa Press, 1994.

Snyder, Leon C. *Gardening in the Upper Midwest.* Second edition. Minneapolis: University of Minnesota Press, 1985.

Snyder, Leon C. *Native Plants for Northern Gardens.* Minneapolis: Anderson Horticultural Library, 1991.

Snyder, Leon C. *Trees and Shrubs for Northern Gardens.* Minneapolis: Anderson Horticultural Library, 2000.

Tekiela, Stan. *Trees of Minnesota: Field Guide.* Cambridge, MN: Adventure Publications, 2001.

Tekiela, Stan. *Wildflowers of Minnesota: Field Guide.* Cambridge, MN: Adventure Publications, 1999.

Tester, John R. *Minnesota's Natural Heritage: An Ecological Perspective.* Minneapolis: University of Minnesota Press, 1995.

Wasowski, Sally. *Gardening with Prairie Plants.* Minneapolis: University of Minnesota Press, 2002.

White, Lee Anne. *Landscaping Your Home: Creative Ideas from America's Best Gardeners.* Newtown, CT: The Taunton Press, 2001.

Wilson, Jim. *Landscaping with Wildflowers: An Environmental Approach to Gardening.* Boston: Houghton Mifflin, 1992.

Wovcha, Daniel S., Barbara C. Delaney, and Gerda E. Nordquist. *Minnesota's St. Croix River Valley and Anoka Sandplain: A Guide to Native Habitats.* Minneapolis: University of Minnesota Press, 1995.

Journals

Wild Ones Journal. Several editions.

Websites

Climatology Working Group: http://www.climate.umn.edu

Minnesota Department of Natural Resources: http://www.dnr.state.mn.us

Minnesota Department of Natural Resources: http://files.dnr.state.mn.us/ecological_services/plant_list9-25-02.pdf

Plant Conservation Alliance's Alien Plant Working Group: http://www.nps.gov/plants/alien

United States Department of Agriculture Plants Database: http://plants.usda.gov

Index

About the
Author

Lynn Steiner

Lynn Steiner's enthusiasm for native plants stems from her childhood curiosity about all things natural. She spent many hours hiking near her childhood home in northeastern Wisconsin, camera and notebook in hand.

Lynn received a bachelor of science in natural resources from the University of Wisconsin. In 1981, she moved to Minnesota and attended the University of Minnesota, where she earned a master's degree in horticulture with a minor in agricultural journalism.

For fifteen years, Lynn was the driving force behind *Northern Gardener*, the official publication of the Minnesota State Horticultural Society. As editor, she took the magazine from a society-based publication to a nationally recognized and respected gardening magazine. Now a freelance writer, editor, and photographer, Lynn's work has appeared in the *Minneapolis StarTribune*, *Gardening How-To*, Sunset's *Midwestern Landscaping Book*, and *Miracle-Gro Plant Care Encyclopedia*.

Lynn lives with her husband, two teenage sons, a golden retriever, and two cats on a turn-of-the-century farmstead in northern Washington County, Minnesota, where she gets great enjoyment from tending her gardens and hunting for native plants in the surrounding countryside.